Progress in IS

More information about this series at http://www.springer.com/series/10440

Christian Czarnecki · Christian Dietze

Reference Architecture for the Telecommunications Industry

Transformation of Strategy, Organization, Processes, Data, and Applications

 Springer

Christian Czarnecki
Düsseldorf
Germany

Christian Dietze
Abu Dhabi
United Arab Emirates

ISSN 2196-8705   ISSN 2196-8713   (electronic)
Progress in IS
ISBN 978-3-319-83578-5   ISBN 978-3-319-46757-3   (eBook)
DOI 10.1007/978-3-319-46757-3

Printed on acid-free paper

This Springer imprint is published by Springer Nature
The registered company is Springer International Publishing AG
The registered company address is: Gewerbestrasse 11, 6330 Cham, Switzerland

Preface

The idea of writing this book came to us while we were together on one of our frequent international business trips in the Middle East. Through our role as consultants, we were involved in a broad variety of project engagements, proposal development activities, presentations, and publications related to the architectural transformation of telecommunications operators. In the last decade, we have supported more than 50 transformational projects in the telecommunications industry worldwide. Most of these dealt with the customization of reference solutions provided by the TM Forum and the alignment of those solutions across different parts of the company. This encouraged us to summarize our experiences in general recommendations and blueprints. Moreover, we have had the good fortune to discuss our viewpoints with executives as well as experts worldwide. We came to the conclusion that there is a significant and indeed increasing demand in this knowledge area for professionals, researchers, and students associated with the telecommunications industry. These facts and our well-founded scientific research experience, combined with practical knowledge from our project engagements across the globe, have motivated us to write this book together.

Preparing the concept of this book as well as the detailed elaboration of the content, while at the same time working as professional consultants and academics, has certainly demanded a lot from both of us. In addition, the fact that each of us has been working on different chapters and sections in different locations and time zones has led to several alignments and iterative updates to ensure the comprehensibility and the consistency of the contents throughout all chapters. In total, it has taken us almost two years to finalize this book.

Writing this book was only possible because of the comprehensive, ongoing support we have received from our project clients and colleagues. Being part of the innovative and international environment at the management consultancy, Detecon International GmbH was one of the lucky circumstances that resulted in the opportunity to summarize our experiences in this book. We would like to express our special appreciation to Issa Nasser Oesterreich and Dr. Kai Grunert, who always supported us and gave us the freedom to realize our ideas on projects worldwide. Furthermore, we had the great pleasure to work with teams of inspiring and knowledgeable colleagues. We were always supported by our colleagues from the MENA office, the International Telco Cluster, and our eTOM knowledge initiative. Whereas it is impossible to list all of their names here, we would like to thank each of them personally. Without the TM Forum, our book would not have been possible, so our sincere thanks go to the whole TM Forum team and the eTOM working group. We would also like to thank our editor Christian Rauscher, who was a great help during the whole publication process, and Patricia Joliet for her excellent work in proofreading our manuscript.

Writing such a book alongside the daily work responsibilities has been a challenge that we were only able to meet through the continuous encouragement of our families and friends. We would like to express our deep gratitude to all of them and especially to Nadine Schultes, Cecilia Carvajal, and Dr. Andreas Dietze.

Düsseldorf   Christian Czarnecki
Abu Dhabi   Christian Dietze
November 2016

Contents

Abbreviations

3G	Third generation (of mobile network technologies)
4G	Fourth generation (of mobile network technologies)
ABE	Aggregated Business Entity
ADM	Architecture Development Method (as a part of TOGAF)
ANSI	American National Standards Institute
ARPU	Average Revenue per User
B2B	Business-to-Business
B2B2C	Business-to-Business-to-Consumer
BPM	Business Process Management
BPMN	Business Process Model and Notation
BSS	Business Support Systems
CCAO	Chief Corporate Affairs Officer
CCO	Chief Commercial Officer
CCTA	Central Computing and Telecommunications Agency
CEM	Customer Experience Management
CEO	Chief Executive Officer
CFO	Chief Financial Officer
CHRO	Chief Human Resources Officer
CIM	Computer-Integrated Manufacturing
CIO	Chief Information Officer
CLCO	Chief Legal and Compliance Officer
CMO	Chief Marketing Officer
CNBO	Chief New Business Officer
COO	Chief Operating Officer
CPO	Chief Procurement Officer
CRM	Customer Relationship Management
CSO	Chief Strategy Officer
CTO	Chief Technology Officer
E2AF	Extended Enterprise Architecture Framework
EAF	Enterprise Architecture Framework

EAM	Enterprise Architecture Management
EPC	Event-driven Process Chain
ERM	Entity Relationship Model
eTOM	enhanced Telecom Operations Map
FEAF	Federal Enterprise Architecture Framework
FTTH	Fiber-to-the-Home
GCC	Gulf Cooperation Council
HIS	Hospital Information Systems
HR	Human Resource (Management)
HTML	Hypertext Markup Language
IaaS	Infrastructure as a Service
IAF	Integrated Architecture Framework
ICT	Information and Communication Technology
IEEE	Institute of Electrical and Electronics Engineers
IFEAD	Institute for Enterprise Architecture Developments
IPTV	Internet Protocol Television
ISP	Internet Service Provider
IT	Information Technology
ITIL	Information Technology Infrastructure Library
ITU	International Telecommunication Union
KPI	Key Performance Indicator
LTE	Long-Term Evolution
M2M	Machine-to-Machine
NGN	Next-Generation Network
NIST	National Institute of Standards and Technology
NOC	Network Operations Center
OECD	Organization for Economic Cooperation and Development
OSI	Open Systems Interconnection
OSS	Operations Support Systems
OTT	Over-The-Top
PaaS	Platform as a Service
PMI	Program Management Institute *or* Post Merger Integration
PRINCE2	PRojects In Controlled Environments, version 2
QoS	Quality of Service
RfP	Request for Proposal
RLC/MAC	Radio Link Control/Medium Access Control
SaaS	Software as a Service
S-BPM	Subject-oriented Business Process Management
SID	Shared Information/Data Model
SIM	Subscriber Identity Module
SIP	Strategy, Infrastructure, and Products (as a part of eTOM)
SME	Small- and Medium-sized Enterprises/Small and Medium Enterprise
SMS	Short Message Service
SOA	Service-Oriented Architecture
TAFIM	Technical Architecture Framework for Information Management

TAM	Telecom Applications Map
TCP/IP	Transmission Control Protocol/Internet Protocol
TEAF	Treasury Enterprise Architecture Framework
TNA	Technology Neutral Architecture
TOGAF	The Open Group Architecture Framework
TV	Television
UML	Unified Modeling Language
VoIP	Voice over Internet Protocol

List of Figures

List of Tables

Chapter 1
Addressing the Transformational Needs of Telecommunications Operators

Abstract The telecommunications industry has changed tremendously during the last decades. Challenges of today's telecommunications operators are, for example, enhanced customer orientation and product innovation combined with cost savings as well as shorter lead times. In many cases, this leads to continuous improvement and restructuring initiatives. Process standardization, automation through new software systems, outsourcing of support activities, and roll-out of new network technologies are just some of the typical topics of these initiatives. In this context, an aligned transformation of organization, processes, applications, data, and network technologies is a key success factor. The overall structure of such transformations is supported by general enterprise architecture methods. From a topical perspective, industry-specific reference solutions are proposed by well-recognized industry organizations, such as the TM Forum. This book explains the whole architectural transformation customized to fit the specific challenges of telecommunications operators. All phases are described, from the planning and set-up to the implementation. While this chapter provides an introduction and summary, in subsequent chapters the following details are discussed: Specifics of the telecommunications industry are described in Chap. 2, methodical principals are explained in Chap. 3, a concrete recommendation for the architecture solution is proposed in Chap. 4, and the planning and implementation are discussed in Chap. 5. This book gives the latest insights into the standard development, shows lessons learned from numerous international projects, and presents well-founded research results. Telecommunication practitioners, enterprise architects, project managers, researchers, and students alike benefit from numerous examples and illustrations.

The telecommunications industry has changed tremendously during the last decades. Challenges of today's telecommunications operators are, for example, enhanced customer orientation and product innovation combined with cost savings as well as shorter lead times. In many cases, this leads to continuous improvement and restructuring initiatives. Process standardization, automation through new software systems, outsourcing of support activities, and roll-out of new network technologies are just some of the typical topics of these initiatives. While some

© Springer International Publishing AG 2017
C. Czarnecki and C. Dietze, *Reference Architecture for the Telecommunications Industry*, Progress in IS, DOI 10.1007/978-3-319-46757-3_1

telecommunications operators are able to plan, design, and implement these changes successfully, others become lost in a labyrinth of unaligned activities. A major challenge is to understand the interrelation between those different topics and to design, plan, and implement a well-defined target picture for the whole enterprise. A typical situation in practice is that, in parallel, the IT department plans the outsourcing of IT services to save operational costs, the marketing department plans the launch of a new IPTV offer to increase revenues, and the technology department plans the harmonization of their production systems. In the end, all these different initiatives might result in minor, local improvements that are paid dearly with various conflicts and difficulties from a cross-functional perspective.

An aligned transformation of organization, processes, applications, and network technologies is a key success factor for today's telecommunications operators. The overall structure of such transformations is supported by general enterprise architecture methods. From a topical perspective, the International Telecommunication Union (ITU) and the TM Forum provide reference solutions, such as the *enhanced Telecom Operations Map* (eTOM) for processes and the *Telecom Applications Map* (TAM) for applications. These reference models are well recognized by the whole value chain of the telecommunications industry and can be seen as de facto standard. However, to gain the full benefits of these standards, a structured approach that shows how to use them in a practical context is essential.

From a practical perspective, the content of this book is particularly beneficial for people working in the telecommunications industry including:

- general top managers;
- managers of IT, network or technology departments;
- process/quality/architecture management departments;
- program management departments;
- project managers and team members of transformation projects;
- consulting companies, freelancers, and system integrators.

In addition, researchers and students receive detailed industry-specific insights into the context of information systems.

1.1 What Is the Structure of This Book?

This book explains the whole architectural transformation customized to fit the specific challenges of telecommunications operators. All phases are described, from the planning and set-up to the implementation (cf. Fig. 1.1). The specifics of the telecommunications industry are described in Chap. 1, and the methodical principals are explained in Chap. 3. Based on these two fundamental topics, a concrete recommendation for the architecture solution is proposed in Chap. 4. This architecture solution combines the general structure of enterprise architectures and reference standards in the telecommunications industry and offers a reference for a

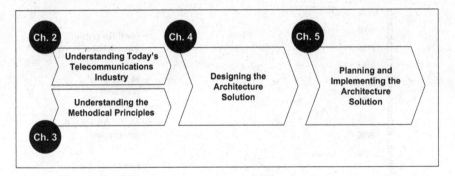

Fig. 1.1 Chapters and their interrelation

concrete solution design. In Chap. 5, the planning and implementation are discussed, using various projects as examples. This book comprises the overarching view of enterprise architecture concepts together with the specific industry standards. It gives the latest insights into the standard development, shows lessons learned from numerous international projects, and presents well-founded research results in enterprise architecture management and reference modeling. Telecommunication practitioners, enterprise architects and project managers alike benefit from numerous examples and illustrations.

1.2 What Are the Major Findings of This Book?

Understanding today's telecommunications industry is a prerequisite for a successful solution design and implementation. The tremendous changes of the industry during the last decades have completely altered their rules and structures. In the past, traditional—mainly government-owned—telecommunications operators were responsible for the technical realization of fixed-line and mobile radio communications. Their business model was based on long-term infrastructure investments that were financed through usage-based connection fees. Today competitors of traditional operators do not necessarily require their own network infrastructure —such as, for example, Over-The-Top (OTT) Providers. Increasingly, the technical connection is becoming a commodity. Innovative applications, convergent services, and dedicated customer orientation are today's success factors. However, increasing data volumes and mobile usage still requires ongoing modernization of network technologies. Figure 1.2 shows the worldwide development of telecommunications subscriptions. During the last 10 years, fixed lines have been in constant decline, while mobile-cellular and especially mobile broadband have increased tremendously.

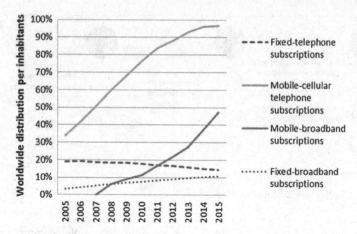

Fig. 1.2 Development of telecommunications subscriptions[1]

Due to the increasing competition, price decreases can be observed (ITU 2015b, p. 5; Plunkett 2014, p. 8). As a result, the revenue growth rates are low: for example, the Telecommunication Industry Association observed an average growth of 6 % in telecommunications spending between 2010 and 2015 Telecommunications Industry Association (2015). A major challenge for telecommunications operators is the combination of continuous innovation requirements with a stagnating market and changing value chains.

Understanding the methodical principles is indispensable for the successful adaptation of structures, processes, and applications to the changed industry conditions. In most cases, those adjustments are related to the various different parts of a telecommunications operator. The planning, design, and realization of those changes are a complex endeavor which, in most situations, takes several years, involves huge project teams, and impacts major parts of the enterprise. Without clear structures and guidelines, the risk of inconsistent and singular solutions is high. The overriding challenge is to understand the interrelations between the different enterprise parts and take decisions that are beneficial from the overall enterprise perspective. The general methical foundation of the solution design is related to information systems modeling. In this context, information systems are a complex construct comprised of employees, their organizational responsibilities, their activities that create the enterprise's outcome, as well as applications that support and automate activities. Enterprise architectures provide a general structure to plan, design, and implement those complex solutions. Content-wise, reference

---

[1]Own illustration, data is based on ITU (2015a). The distribution per inhabitants is a theoretical figure based on the total number of subscriptions and the world population. It does not provide penetration rate—i.e., through the high mobile penetration in developed countries, one mobile subscriber often has several subscriptions. The number of worldwide mobile-cellular telephone subscriptions in comparison to the world population has almost reached the 100 % mark.

models are used as recommendations. In the telecommunications industry, the TM Forum offers well-accepted reference models for processes, data, and applications. From the dynamic perspective, concepts of enterprise architecture management and enterprise transformation support the planning and implementation.

Designing the architecture solution combines the methodical principles in an architectural construct that offers clear recommendations for the specific challenges facing today's telecommunications operators. First, the relevant elements are identified and arranged in an architecture structure for organization, processes, data, and applications. As an additional structural element, five industry-specific architecture domains are proposed. These architecture domains provide an overall structure of telecommunications operators. The customer-centric domain covers all architecture elements related to direct customer interactions. All technical specifics are encapsulated in the technology domain. The product domain includes the planning, development, and roll-out of new products. Both the product and the

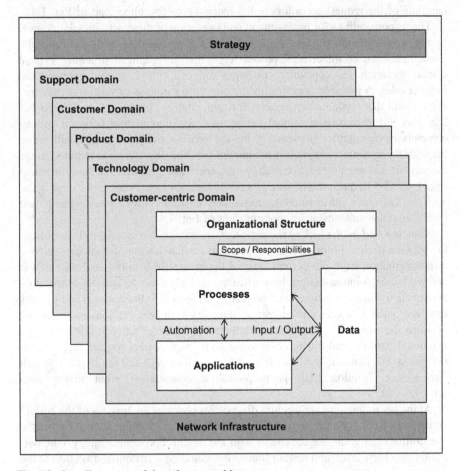

Fig. 1.3 Overall structure of the reference architecture

technology domain prepare the prerequisites to fulfill customer requests in the customer-centric domain. Further support activities are included in the customer domain and enterprise support domain. For each of these domains, concrete reference solutions for organization, processes, data, and applications are described (cf. Fig. 1.3). These reference solutions combine the industry-specific TM Forum reference models and provide a detailed blueprint for the transformational needs of telecommunications operators. The reference architecture includes a hierarchical decomposition and interrelations between the different elements.

In the following, an exemplary description of the different elements of the reference architecture proposed in this book is given (cf. Fig. 1.4). The customer-centric domain contains seven reference process flows defining all interactions with a customer from an end-to-end perspective. The Request-to-Answer process is one of these reference process flows. It deals with answering all types of customer requests. The process can be divided into the following activities: customer contact management, request specification, and the handling of the request according to the request type (cf. upper part of Fig. 1.4).

The responsibilities for management and execution of these activities are defined by the organizational structure. Parameters for structuring those responsibilities are contact channel, customer type, product type, and geographical structure. Typical contact channels for consumer customers are call center, shops, internet, and indirect sales. A possible organizational structure is a consumer sales and customer service unit that contains departments for each contact channel (cf. middle part of Fig. 1.4). From an organizational perspective, a differentiation between contact channels is reasonable. However, from the process perspective, standardization between those contact channels is recommended. The data elements required in the Request-to-Answer process are mainly customers and products (cf. bottom left of Fig. 1.4). The Request-to-Answer process is mapped to various application areas, such as customer information management, customer order management, and customer self-management (cf. bottom right of Fig. 1.4).

Planning and implementing the architecture solution is essential to benefit from the solution design. From a dynamic perspective the architectural implementation is a transformation from the current state of the enterprise to a targeted state that is defined by the solution design. In most cases, the entire design and implementation are conducted in a cross-functional project. With respect to the duration and persons involved, such a project can be seen as complex endeavor. Various interrelations between the architectural elements, conflicts of objective between different organizational entities, and changing external or internal factors require careful consideration. For planning the tasks from the set-up to design and implementation, an Architecture Solution Map is proposed. It consists of eight major tasks (cf. Fig. 1.5).

At the beginning, the architecture diagnostics provides an analysis of the current situation as basis for a first goal definition. The strategic alignment ensures that the transformational goals are consistent with the overall corporate strategy. The definition of a high-level architecture framework could be a customized version of the reference architecture proposed in this book. Typically a cross-functional

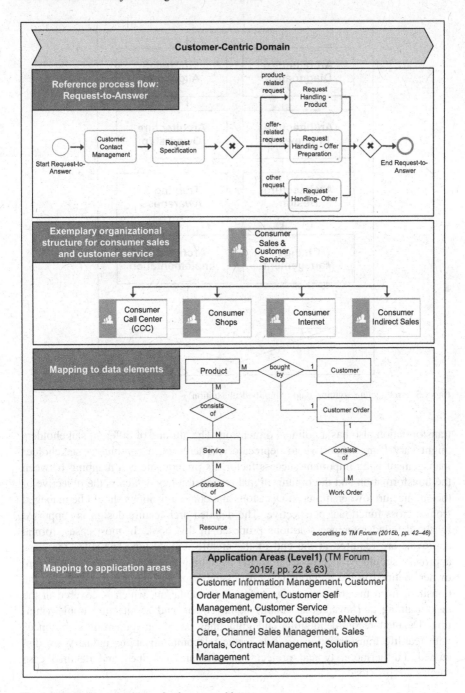

Fig. 1.4 Exemplary detailing of reference architecture

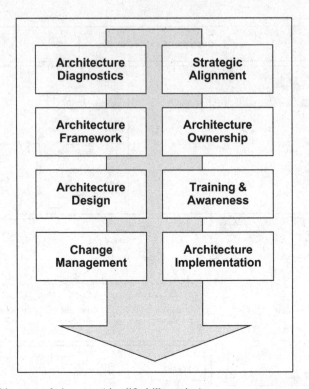

Fig. 1.5 Architecture solution map (simplified illustration)

transformation also has a political dimension. The attitude of different stakeholders might vary from supportive to deprecatory. Therefore, a continuous stakeholder management is an important success factor. Its prerequisite is a mapping between the transformation and the organizational responsibilities, which is the objective of the architecture ownership task. Decisions and communications should be managed from a cross-functional perspective. The detailed architecture design is supported by the detailed reference solutions proposed in this book. In most cases, formal approvals are required at the end of the architecture design. The rules for these approvals should be clearly set from the beginning and linked to the architecture owners defined in the previous task. An essential part of the transformation is the transition from the solution design to its implementation, which is covered in the tasks training and awareness, change management, and architecture implementation. Detailed recommendations and guidelines based on numerous experiences with real-life transformation projects in the telecommunications industry are discussed. Furthermore, typical project examples are described and detailed case studies are provided.

1.3 Which Sources Were Used?

The content provided in this book is based on practical and scientific work. Both authors have broad experience in an international consulting company specialized on the telecommunications industry. They were involved in more than 50 transformation projects of telecommunications operators worldwide. Since 2006 they have been active members in the TM Forum, and have had leading roles in the further development of eTOM. As part of consulting projects and topical development, they have developed various reference solutions that have helped telecommunications operators worldwide. This book summarizes the practical experience with real-life transformation projects and insights from the telecommunications industry.

Furthermore, the content is based on many years of scientific research. Results have been discussed at international conferences and published in well-recognized journals (cf. Table 1.1).

This book follows the design science paradigm (e.g. Hevner et al. 2004; Peffers et al. 2007) that can be roughly divided into problem identification, solution design, and evaluation. While a concrete model has to fit to a single situation, the artifacts described in this book are a point of reference for a wide range of situations.[2] The content is therefore based on the generalized solutions of various transformation projects and the work of the industry organization TM Forum. The focus of this

Table 1.1 Published own research results

References	Topic
Czarnecki (2009)	Discussion of inter-company customer relationship management in the telecommunications industry
Czarnecki et al. (2009), Czarnecki and Spiliopoulou (2012)	Design of an enterprise architecture framework for a Next Generation Network (NGN)
Czarnecki et al. (2010)	Design of a reference framework for process virtualization in the telecommunications industry
Czarnecki et al. (2011)	Analysis of customer orientation based on real-life transformation projects
Czarnecki et al. (2012)	Identification of project types based on an analysis of real-life transformation projects
Czarnecki et al. (2013)	Extension of eTOM by reference process flows
Czarnecki (2013)	Design of a reference architecture for telecommunications operators[a]

[a]Starting point of this book are the results published in Czarnecki (2013). They are presented in this book in a translated, completely revised, and substantially extended version

---

[2]Please see e.g. Fettke and Loos (2007a, p. 4) and Sect. 3.3 for further details on the differentiation between concrete and reference models.

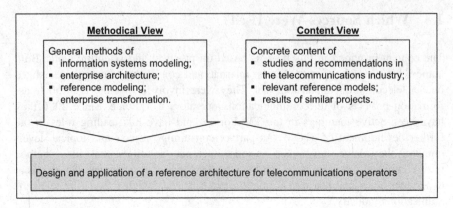

Fig. 1.6 Relevant topics

book is a detailed description of the resulting reference artifacts. In this context, artifacts are considered as a general object resulting from a design activity (Hevner et al. 2004). Please see Czarnecki et al. (2010, 2013) as well as Czarnecki and Spiliopoulou (2012) regarding details about the scientific design and evaluation process.

The design and application of a reference architecture for the telecommunications industry can be structured into the methodical and content perspective (cf. Fig. 1.6).

From the methodical perspective, information systems modeling (e.g. Avison and Fitzgerald 2006; Satzinger 2015; Stair and Reynolds 2012), enterprise architecture (e.g. Ahlemann 2012; Schekkerman 2004; Van Den Berg and Van Steenbergen 2006; Winter and Sinz 2007), reference modeling (Becker et al. 2003; e.g. Becker and Delfmann 2007; Fettke and Loos 2007b; Thomas 2006), and enterprise transformation (e.g. Aier and Gleichauf 2010; Aier and Weiss 2012; Alt and Zerndt 2009; Jetter et al. 2009; Young and Johnston 2003) are relevant topics. Those concepts provide general guidelines that are irrespective of the concrete problem domain of the telecommunications industry. They help to identify and structure the relevant parts and their interrelations. The content is derived from industry-specific studies and recommendations (e.g. Copeland 2009; Grishunin and Suloeva 2015; Grover and Saeed 2003; Misra 2004; Plunkett 2014; Yahia et al. 2006), relevant reference models (e.g. ITU 2007; Kelly 2003; Orand 2013; TM Forum 2015), and project results. A detailed discussion of those topics and related literature is provided in Chaps. 1 and 3.

1.4   Who Could Benefit from This Book?

The content of this book has a practical relevance for managers, experts, developers, researchers, and students associated with the telecommunications industry. It covers organizations, processes, data, and applications for all functional parts of a telecommunications operator. The provided solutions are structured in a hierarchical manner from high-level frameworks to more detailed solutions. The proposed solutions are combined with practical examples, guidelines and recommendations, thereby providing helpful content from business and technical as well as management and operational perspectives. However, the focus of this book is on the conceptual information systems modeling—i.e., concrete software solutions or technical infrastructure specifications are not included. With its scientific foundation, the book can be used by researchers and students either for concrete work related to the telecommunications industry or as a domain-specific example in the field of information systems modeling.

For employees involved with telecommunications operators, exemplary usage scenarios of this book are:

- *General top managers* gain an overview regarding today's industry challenges and the related parts of their enterprise (cf. Chap. 1). They can utilize the high-level architecture framework (cf. Sect. 4.1) and the typical project examples (cf. Sects. 5.3 and 5.4) to define strategic transformation initiatives.
- *Managers of IT, network or technology departments* are typically directly involved in the realization of technical changes (e.g. new software systems or roll-out of network infrastructure). They can use the reference architecture (cf. Chap. 4) for an end-to-end understanding and identification of possible inter-relations between those technical changes and business requirements.
- *Process/quality/architecture management departments* can use the methodical basics (cf. Chap. 3) as well as the reference solutions (cf. Chap. 4) as blueprints for their concrete solution design.
- *Program management departments* are typically responsible for the set-up of strategic projects as well as the identification of overlaps and synergies. This tasks are supported by the industry challenges (cf. Chap. 1), the high-level architecture framework (cf. Sect. 4.1), as well as the recommendations for planning and implementing the architecture solution (cf. Chap. 5).
- *Project managers and team members of transformation projects* are directly involved in the planning, design, and implementation of the architectural elements described in this book. They can use the content in this book as methodical guidelines (cf. Chap. 3) and as concrete blueprints for their solution design (cf. Chap. 4). Furthermore, the insights from the industry (cf. Chap. 1) and real-life projects (cf. Chap. 5) support them in their project work.

- *Consulting companies, freelancers, and system integrators* might be involved in different parts of a transformation project. Based on their role, they can use this book either for a high-level overview of industry-specific requirements (cf. Chap. 1 and Sect. 4.1), concrete solution blueprint (cf. Chap. 4) or lessons learned from similar projects (cf. Chap. 5).

Researchers and students might use this book as follows:

- *Industry-specific artifacts* are common content in research. The telecommunications industry is a widespread research topic (e.g. Bruce et al. 2008; Grover and Saeed 2003; Mikkonen et al. 2008).[3] This book provides specific results, insights, and examples of the telecommunications industry that can be used as input or starting point by researchers.
- *Information systems modeling* is an important field in various study courses. At some point, concrete practical examples are required in order to explain or apply general concepts. This book provides concrete examples of the telecommunications industry for different model types.
- *Reference modeling* is an important topic of information systems research and teaching. Reference models provide generalized solutions for their re-use in similar situations. They are therefore, in most cases, domain-specific (e.g. Fettke and Loos 2007a, p. 4).[4] This book provides insights and examples of well-accepted reference models as well as their concrete application.

1.5 How Is the Content Used in Real-Life Projects?

The proposed reference architecture provides a generalized structure and recommendation for telecommunications operators. Its usage in a concrete project requires customization and depends on the specific project scope. The level of change intended by the project might vary from a documentation or analysis of the current situation to a complete reengineering. The relevant parts of the enterprise might range from well-defined enterprise units (e.g. a call center for consumer sales) to cross-functional topics impacting various enterprise parts (e.g. the CRM system). The authors have used the described reference architecture in various real-life projects. In this book four of these projects are described as case studies (cf. Table 1.2). Further details about these case studies are described in Sect. 5.4.

---

[3]Please see Chap. 2 for further references on research in telecommunications industry.
[4]Please see Sect. 3.3 for further details on reference modeling.

Table 1.2 Case studies of real-life projects

Project	Scope	Usage of reference architecture
Introduction of an OSS	Development of a concept for introducing an OSS as part of a transformation program	• Reference process flows as blueprint • Assignment of processes to application functions for the selection of concrete software solutions • Overarching data model as integration element
Introduction of a CRM system	Selection and introduction of a standardized CRM system that is realized through customized standard software	• Reference process flows for the customer-centric domain • Requirements engineering through mapping of processes to application functions • Overarching data model as integrating element
Introduction of an OSS	Development of an overall architecture for the introduction of an OSS based on an NGN	• Reference process flows of the technology domain for requirements engineering • Reference process flows of the customer-centric, customer and product domain for integration • Focus on OSS, usage of functions for structuring • Overarching data model as integrating element
Introduction of process architecture	Introduction of an overarching process architecture as basis for a company-wide management of business processes	• Reference process flows of all five domains as blueprint • Process improvements for two reference process flows of the customer-centric domain • Mapping of existing software systems through the application functions • Overarching data model as requirement

References

Ahlemann, F. (Ed.). (2012). *Strategic enterprise architecture management: Challenges, best practices, and future developments, management for professionals*. Berlin; New York: Springer.

Aier, S., & Gleichauf, B. (2010). Towards a systematic approach for capturing dynamic transformation in enterprise models. In *Forty-Third Annual Hawaii International Conference on System Sciences (HICSS-43)* (pp. 1–10). IEEE Computer Society.

Aier, S., & Weiss, S. (2012). Facilitating enterprise transformation through legitimacy—An institutional perspective. In D. C. Mattfeld & S. Robra-Bissantz (Eds.), *Multi-Konferenz Wirtschaftsinformatik 2012—Tagungsband Der MKWI 2012* (pp. 1073–1084). Braunschweig: GITO Verlag.

Alt, R., & Zerndt, T. (2009). Grundlagen der transformation. In R. Alt, B. Bernet, & T. Zerndt (Eds.), *Transformation von Banken Praxis des In-und outsourcings in der finanzindustrie* (pp. 47–68). Berlin: Springer.

Avison, D., & Fitzgerald, G. (2006). *Information systems development: Methodologies, techniques & tools,* (4th ed.). London: McGraw-Hill.

Becker, J., & Delfmann, P. (2007). *Reference modeling efficient information systems design through reuse of information models.* Heidelberg: Physica Verlag.

Becker, J., Kugeler, M., & Rosemann, M. (2003). *Process management a guide for the design of business processes.* Berlin: Springer.

Bruce, G., Naughton, B., Trew, D., Parsons, M., & Robson, P. (2008). Streamlining the telco production line. *Journal of Telecommunications Management, 1,* 15–32.

Copeland, R. (2009). *Converging NGN wireline and mobile 3G networks with IMS.* Boca Raton: CRC Press.

Czarnecki, C. (2009). Gestaltung von customer relationship management über die Grenzen von Telekommunikationsunternehmen hinweg. In T. Eymann (Ed.), *Bayreuther Arbeitspapiere Zur Wirtschaftsinformatik—Tagungsband Zum Doctoral Consortium Der WI 2009* (pp. 11–23). Bayreuth: Universität Bayreuth.

Czarnecki, C. (2013). *Entwicklung einer referenzmodellbasierten Unternehmensarchitektur für die Telekommunikationsindustrie.* Berlin: Logos-Verl.

Czarnecki, C., & Spiliopoulou, M. (2012). A holistic framework for the implementation of a next generation network. *International Journal of Business Information Systems, 9,* 385–401.

Czarnecki, C., Heuser, M., & Spiliopoulou, M. (2009). How does the implementation of a next generation network influence a telecommunication company? In *Proceedings of the European and Mediterranean Conference on Information Systems (EMCIS).* Izmir, Turkey: Brunel University.

Czarnecki, C., Winkelmann, A., & Spiliopoulou, M. (2010). Services in electronic telecommunication markets: A framework for planning the virtualization of processes. *Electronic Markets, 20,* 197–207.

Czarnecki, C., Winkelmann, A., & Spiliopoulou, M. (2011). Making business systems in the telecommunication industry more customer-oriented. In J. Pokorny, V. Repa, K. Richta, W. Wojtkowski, H. Linger, C. Barry, & M. Lang (Eds.), *Information systems development* (pp. 169–180). New York: Springer.

Czarnecki, C., Winkelmann, A., & Spiliopoulou, M. (2012). Transformation in telecommunication—Analyse und clustering von real-life projekten. In D. C. Mattfeld & S. Robra-Bissantz (Eds.), *Multi-Konferenz Wirtschaftsinformatik 2012—Tagungsband Der MKWI 2012* (pp. 985–998). Braunschweig: GITO Verlag.

Czarnecki, C., Winkelmann, A., & Spiliopoulou, M. (2013). Reference process flows for telecommunication companies: An extension of the eTOM model. *Business & Information Systems Engineering, 5,* 83–96. doi:10.1007/s12599-013-0250-z

Fettke, P., & Loos, P. (2007a). Perspectives on reference modeling. In P. Fettke & P. Loos (Eds.), *Reference modeling for business systems analysis* (pp. 1–21). Hershey: IGI Global.

Fettke, P., & Loos, P. (Eds.). (2007b). *Reference modeling for business systems analysis.* Hershey, PA: Idea Group Pub.

Grishunin, S., & Suloeva, S. (2015). Project controlling in telecommunication industry. In S. Balandin, S. Andreev, & Y. Koucheryavy (Eds.), *Internet of things, smart spaces, and next generation networks and systems* (pp. 573–584). Cham: Springer International Publishing.

Grover, V., & Saeed, K. (2003). The telecommunication industry revisited. *Communications of the ACM, 46,* 119–125. doi:10.1145/792704.792709

Hevner, A. R., March, S. T., Park, J., & Ram, S. (2004). Design science in information systems research. *MIS Quarterly, 28,* 75–105.

ITU. (2007). *ITU-T recommendation M.3050.1: Enhanced telecom operations map (eTOM)—The business process framework.*

ITU. (2015a). *Key ICT indicators for developed and developing countries and the world.*

ITU. (2015b). *ICT facts and figures—The world in 2015.*

Jetter, M., Satzger, G., & Neus, A. (2009). Technological innovation and its impact on business model, organization and corporate culture—IBM's transformation into a globally integrated, service-oriented enterprise. *Business & Information Systems Engineering, 1*, 37–45. doi:10.1007/s12599-008-0002-7

Kelly, M. B. (2003). The telemanagement forum's enhanced telecom operations map (eTOM). *Journal of Network and Systems Management, 11*, 109–119.

Mikkonen, K., Hallikas, J., & Pynnönen, M. (2008). Connecting customer requirements into the multi-play business model. *Journal of Telecommunications Management, 2*, 177–188.

Misra, K. (2004). *OSS for telecom networks: An introduction to network management.* London: Springer.

Orand, B. (2013). *Foundations of IT service management: With ITIL 2011.*

Peffers, K., Tuunanen, T., Rothenberger, M. A., & Chatterjee, S. (2007). A design science research methodology for information systems research. *Journal of management information systems, 24*, 45–77. doi:10.2753/MIS0742-1222240302

Plunkett, J. W. (2014). *Plunkett's telecommunications industry almanac 2015: The only comprehensive guide to the telecommunications industry.*

Satzinger, J. W. (2015). *Systems analysis and design in a changing world* (7th ed.). Boston, MA: Cengage Learning.

Schekkerman, J. (2004). *How to survive in the jungle of enterprise architecture frameworks: Creating or choosing an enterprise architecture framework.* Victoria: Trafford.

Stair, R. M., & Reynolds, G. W. (2012). *Fundamentals of information systems.* Boston: Course Technology/Cengage Learning.

Telecommunications Industry Association. (2015). *TIA's 2015–2018 ICT market review & forecast.*

Thomas, O. (2006). Understanding the term reference model in information systems research: History, literature analysis and explanation. In C. Bussler & A. Haller (Eds.), *Business process management workshops* (pp. 484–496)., Lecture Notes in Computer Science Berlin, Heidelberg: Springer.

TM Forum. (2015). *Business process framework (eTOM): Concepts and principles (GB921 CP).* Version 15.0.0. ed.

Van Den Berg, M., & Van Steenbergen, M. (2006). *Building an enterprise architecture practice.* Netherlands, Dordrecht: Springer.

Winter, R., & Sinz, E. J. (2007). Enterprise architecture. *Information Systems and e-Business Management, 5*, 357–358. doi:10.1007/s10257-007-0054-0

Yahia, I. G. B., Bertin, E., & Crespi, N. (2006). Next/new generation networks services and management. In *Proceedings of the International Conference on Networking and Services, ICNS'06* (p. 15). Washington, DC, USA: IEEE Computer Society. doi:10.1109/ICNS.2006.77

Young, L. W., & Johnston, R. B. (2003). The role of the Internet in business-to-business network transformations: A novel case and theoretical analysis. *Information Systems and e-Business Management, 1*, 73–91. doi:10.1007/BF02683511

Chapter 2
Understanding Today's Telecommunications Industry

Abstract Understanding today's telecommunications industry is a prerequisite for a successful architectural transformation. The tremendous changes of the industry during the last decades have completely altered their rules and structures. In the past, traditional—mainly government-owned—telecommunications operators were responsible for the technical realization of fixed-line and mobile radio communications. Their business model was based on long-term infrastructure investments that were financed through usage-based connection fees. Today, competitors of traditional operators do not necessarily require their own network infrastructure—such as, for example, Over-The-Top (OTT) providers. Increasingly, the technical connection is becoming a commodity. Innovative applications, convergent services, and dedicated customer orientation are today's success factors. However, increasing data volumes and mobile usage still requires ongoing modernization of network technologies. A major challenge for telecommunications operators is the combination of continuous innovation requirements with a stagnating market and changing value chains. Section 2.1 explains the market conditions and ecosystem with respect to price decrease and cost pressure, competition through Over-the-Top providers, new opportunities in vertical markets, and challenges for regulators. The interrelation between commercial and technical products as well as changed customer demands and usage behavior are discussed in Sect. 2.2. The value chain reacts to the changed market conditions through increased fragmentation of the value creation and new partnering, which are topics of Sect. 2.3.

The telecommunications industry is currently going through a major transformation which creates both opportunities and challenges for fixed operators, mobile operators as well as Internet service providers (e.g. Grover and Saeed 2003; Picot 2006; Plunkett 2014). New and innovative players are entering the telecommunications market, and this has led to a restructuring of the whole telecommunications industry (Pousttchi and Hufenbach 2011; Wulf and Zarnekow 2011a). Through the fast technological development, increasing market dynamics and deregulation in many countries, the complexity in the telecommunications industry is constantly increasing (Plunkett 2014, pp. 7–9).

© Springer International Publishing AG 2017
C. Czarnecki and C. Dietze, *Reference Architecture for the Telecommunications Industry*, Progress in IS,
DOI 10.1007/978-3-319-46757-3_2

Those changes and challenges of the telecommunications industry are the topic of various publications and studies with different focus, including overall market research (Plunkett 2014), value creation and market players (Grover and Saeed 2003; Peppard and Rylander 2006; Picot 2006; Pousttchi and Hufenbach 2011; Tardiff 2007; Wulf and Zarnekow 2011a), (de)regulation and competition (Cave et al. 2002; Gentzoglanis and Henten 2010), standardization (Lyall 2011), structures and processes (Bruce et al. 2008; Czarnecki et al. 2013; Pospischil 1993) as well as various functional or technical specifics (e.g. Copeland 2009; Czarnecki and Spiliopoulou 2012; Grishunin and Suloeva 2015; Lewis 2001; Mikkonen et al. 2008; Misra 2004; Yahia et al. 2006).

The first challenge of today's telecommunications industry is to understand the various players. In the past, the technical realization of communication via mobile or fixed-line networks was the major objective of telecommunications operators.[1] The convergence of voice, video, and data has led to mergers, acquisitions, and partnerships (Tardiff 2007, p. 132; Wulf and Zarnekow 2011b, pp. 10–11). Increasingly, application and content offers are intermixed with telecommunication services (Peppard and Rylander 2006, pp. 133–134). Entertainment services such as TV offers are linked to traditional communication services, resulting in new competition between TV cable operators and communication network operators (Plunkett 2014, p. 7). The convergence of telecommunications, media, and hardware industries is an already observed implication (Arlandis and Ciriani 2010, p. 121).

Plunkett (2014, pp. 7–8) points out that the exact composition of the telecommunications industry varies when it comes to including or excluding certain business sectors—e.g., communication equipment or related consulting services. Arlandis and Ciriani (2010, pp. 121–124) relate the telecommunications industry to the information and communication technology (ICT) sector, which they define as an ecosystem consisting of technologies providers, network operators, platform operators, and content providers. Grover and Saed (2003, p. 120) propose a categorization of the telecommunications industry into network providers, tool providers, transaction/service providers, and internet/content providers.

When it comes to concrete enterprises offering telecommunication products and services, there is a huge range of different business models, including branded resellers, mobile virtual network operators, or mobile virtual network enablers (Pousttchi and Hufenbach 2009, p. 87). There is a variety of characteristics to differentiate those business models—e.g., functional coverage of the value chain or level of control of the communication network (Kimiloglu et al. 2011, pp. 40–41). A clear understanding of the market positioning and business scope of a

---

[1]In this book the term *telecommunications operator* is used for all firms offering, providing, and operating telecommunication products and services. It can be seen as synonym for *telecommunication company* or *telecommunication firm*. It is understood as a generic term including, e.g., *telephone company or communication service provider*. A telecommunications operator might offer different telecommunications services (e.g., voice or data) to different customer segments (e.g., residential or wholesale).

Customer	consumer		business (retail)	business (wholesale)
Value Chain	component	subsystem	network system	device
	network		service	content/application
Business Activities	production	operation & maintenance	sales	after-sales
Network	fixed line		mobile	satellite

Fig. 2.1 Framework for categorizing telecommunications operators (according to Czarnecki 2013, p. 52)[2]

telecommunications operator is an essential prerequisite to support its transformational needs. Therefore, in this book, a categorization along the dimensions *customer*, *value chain*, *business activities*, and *network* is proposed (cf. Fig. 2.1). The different dimensions and characteristics are based on a review of existing categorization criteria related to the telecommunications industry (Cave et al. 2002; Doeblin and Dowling 2007; Ehrmann 1999; Fransman 2002, p. 475; Gerpott 2003, p. 1090; ITU 1998, p. 13; Maitland et al. 2002; Picot 2006; Pousttchi and Hufenbach 2011).

The dimension *customer* specifies the intended end customer(s) of the telecommunications operator. It is differentiated into consumer, business (retail), and business (wholesale). The *value chain* starts with the technical hardware and software prerequisites of communication networks (component, subsystem, network system, and device). The network covers all technical aspects required to realize services which might be related to content or applications. The *business activities* are divided into production, operations and maintenance, sales, and after-sales. The *network* can be specified by fixed line, mobile, and satellite. The scope of a concrete telecommunications operator might be a complex mixture of the above characteristics.

Telecommunications operators are confronted with various challenges that influence their transformational needs. Those challenges are summarized along the dimensions *market*, *products/services*, and *value chain* (cf. Fig. 2.2).

The *market* conditions have changed due to convergence that leads to increased competition (Cave et al. 2002; Plunkett 2014, pp. 7–22; Wulf and Zarnekow 2011a, pp. 290–292). Those changes of the market structures and ecosystem (Arlandis and Ciriani 2010, pp. 124–129) result in new market potentials (Basole and Karla 2011, pp. 313–314; Kimiloglu et al. 2011, pp. 47–48; Pousttchi and Hufenbach 2009, p. 87) combined with increased cost and price pressure. Furthermore, these changes

---

[2]Translated and revised version of the illustration published in Czarnecki (2013, p. 52).

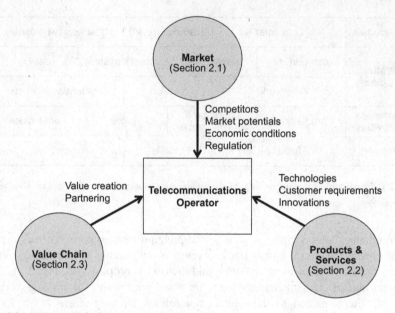

Fig. 2.2 Challenges of telecommunications operators

lead to new requirements and challenges for regulators (Tardiff 2007). The *value chain* reacts to the changed market conditions through increased fragmentation of the value creation (Peppard and Rylander 2006, pp. 128–129; Pousttchi and Hufenbach 2011, p. 307) and new partnering (Grover and Saeed 2003, pp. 121–125). In the dimension *products and services*, telecommunications operators are confronted with the complexity of production systems (Bruce et al. 2008; Misra 2004) as well as changed customer demands and usage behavior (Gans et al. 2005, pp. 256–259; Taylor 2002, pp. 126–135). Both are related to the requirement of continuous innovations (Picot 2006) and shorter product development cycles (Bruce et al. 2008). Those challenges are an important factor for the transformation of telecommunications operators. Therefore, they are further discussed in the following sections: telecommunications market in Sect. 2.1, telecommunications products and services in Sect. 2.1.3, and telecommunications value chains in Sect. 2.2.2.

2.1 Telecommunications Market

The telecommunications market has changed tremendously. The resulting cost and price pressure and their impact on telecommunications operators are discussed in Sect. 2.1.1. Convergence leads to increased competition through Over-the-Top (OTT) providers that offer content and application services on top of existing communication services. The challenges of OTT providers for traditional telecommunications operators are summarized in Sect. 2.1.2. In summary, the changed market conditions lead to the disappearance of former revenue sources. New revenue potentials could be realized in vertical markets, which are discussed in Sect. 2.1.3. Furthermore, these changes result in new requirements and challenges for regulators (Tardiff 2007) as illustrated in Sect. 2.1.4.

2.1.1 Price Decrease and Cost Pressure

From an economic perspective, the telecommunications industry is an important part of the ICT sector. Global revenue figures are provided by various analysts and research companies. They depend on the exact definition of the industry being applied for their calculation. Plunkett (2014, p. 8) uses a broad definition and estimated a global revenue of 5.4 trillion USD for 2014. The Telecommunications Industry Association (2015) publishes a global revenue of 5.6 trillion USD. Bloomberg[3] defines *Telecom Carriers* as an own industry with a total revenue of 2.1 trillion USD. When it comes to the future trend, these analysts and research companies forecast a slight revenue growth for the next years. However, this revenue growth is decreasing. From a global perspective, the telecommunications industry is a stagnating market.[4]

For a differentiated understanding of the telecommunications industry, the following figures should be considered:

- The worldwide number of fixed-telephone subscriptions has been declining since 2006, from 1.26 to 1.10 billion in 2014 (ITU 2015a).
- The worldwide number of mobile-cellular telephone subscriptions has more than doubled since 2006, from 2.75 to 6.95 billion in 2014. However, the growth rate is decreasing (ITU 2015a).
- The worldwide number of broadband subscriptions (fixed and mobile) is increasing. Mobile-broadband subscriptions have especially demonstrated a tremendous growth, from 0.27 billion in 2007[5] to 2.69 billion in 2014 (ITU 2015a).

---

[3]Bloomberg offers an online tool called *Bloomberg Industry Market Leaders* (*Visual Data*) that provides key metrics of 49 industries and 580 leading companies (please see www.bloomberg.com/visual-data/industries/). The figure cited here was accessed in Dec. 2015.

[4]From the macroeconomic perspective the access to modern telecommunication infrastructure is a critical success factor for economic growth and wealth. Please see, e.g., Hanna (2010), Laudon and Traver (2015), and OECD (2014) for further information. This book focuses on the microeconomic perspective—i.e., the impact of the changed conditions for telecommunications operators.

[5]The mobile-broadband technology started with 4G in 2006 (Plunkett 2014, p. 495). Therefore, ITU provides figures for mobile-broadband subscriptions from 2007 onwards. (Plunkett 2014, p. 495).

- The market penetration for communication services is constantly increasing: the global estimates for 2015 by ITU (2015b, pp. 2–3) are 69 % of 3G population coverage, 46 % of households with internet access, and 46 % of individuals with mobile-broadband subscriptions. For the member countries of the *Organisation for Economic Co-operation and Development* (OECD), the penetration is much higher, with an estimated 81 % for mobile-broadband subscriptions (OECD 2015).
- For most communication services, a price decrease can be observed (ITU 2015b, p. 5; Plunkett 2014, p. 8) which is a result of the increased competition and ongoing deregulation of the market. For example, ITU (2015b, p. 5) shows decreasing prices for fixed-broadband between 2008 and 2011 with a stagnation since then.

Telecommunications operators are confronted with tremendous changes in the usage behavior in a stagnating market—e.g., compared to a basic mobile phone, using a smartphone generates more than 14 times the data volume (Verma and Verma 2014). This growth of the data volume has to be handled under the condition of stagnating or even decreasing prices. In the past, traditional communication services—for example, voice telephony—were the major revenue sources of telecommunications operators. Now, the pure transmission is becoming more and more of a commodity for the customer. The increasing demand for high transmission bandwidths still requires extensive investments in network infrastructure. However, those same networks are then beneficial for content and application providers such as Google, Facebook, and Netflix, that can profit from the resulting revenues without any participation in the infrastructure investment. For further information please see the discussion about *net neutrality* (e.g. Belli and De Filippi 2015; Plunkett 2014, p. 10). Furthermore, those content and application providers even compete with traditional telecommunications operators. As a result, telecommunications operators require innovative services to secure their revenues. Hence, the two contrary conditions of a stagnating and innovative market are mixed. For telecommunications operators, this means the combination of cost reduction and efficiency increase in order to realize the financial flexibility for investments in innovative services.

This financial situation is further complicated through new competition caused by the convergence of the market. The technical capability for a broadband transmission requires major investments in fixed or mobile network infrastructure. The value proposition recognized by the customer is related to the communication service. And today those communication services can be offered without owning any network infrastructure. For example, the launch of smartphones—which was seen by the telecommunications operators as an opportunity to introduce new services leading to higher *Average Revenue Per User* (ARPU)—has actually been a facilitator for the introduction of new services by Over-the-Top (OTT) providers (cf. Sect. 2.1.2). The new services offered by OTT providers have replaced equivalent telecommunication services—e.g., *WhatsApp* in the messaging market has replaced the traditional Short Messaging Service (SMS).

In the voice market, IP-based products such as *Skype* and other highly complex enterprise applications have resulted in falling revenues for telecommunications operators. In fact, the usage of Voice over Internet Protocol (VoIP) is massively changing the telecommunications industry (Plunkett 2014, pp. 14–16). As a consequence, the traditional voice and messaging markets for telecommunications operators are constantly in decline. A significant part of both historic and predicted telephony and messaging market shifts can be attributed to regulation—either directly related to pricing (e.g., changes in maximum termination or roaming fees), or through the introduction of more competition (e.g., new licensees and wholesale rules). Section 2.3.2 provides a more detailed look at the new role of regulators in today's telecommunications industry.

For telecommunications operators, the changed market conditions require higher efficiency and flexibility. In most cases, this leads to transformations of operational structures. These transformations are supported by the reference architecture described in this book. From a strategic perspective, telecommunications operators have to combine their technical capabilities with revenue to create new value propositions. For integrated telecommunications operators—i.e., those operating fixed and mobile network infrastructures—a strategic option is the bundling of communication services and enrichment with content. A typical example is a quadruple-play service combining mobile and fixed telephony, broadband internet, and IPTV. In most cases, this requires partnering with content providers (Grover and Saeed 2003, pp. 121–125). With those product bundles, telecommunications operators enter the television, video, and media markets. The results are new competitors, such as television cable companies,[6] (Plunkett 2014, p. 17) and increased complexity of the value creation (Peppard and Rylander 2006, pp. 128–129; Pousttchi and Hufenbach 2011, p. 307). Moreover, those services require a high bandwidth. Therefore, increasing the bandwidth of the offered data connection is an additional strategic option. As example, launching Fiber-to-the-Home (FTTH) services is currently an important topic for telecommunications operators (Plunkett 2014, pp. 17–18).

In summary, from a financial perspective telecommunications operators are confronted with price decrease and cost pressure. Both are related to changed usage behaviors and strong competition in convergent markets. In response, telecommunications operators have to realize new revenue sources through innovative services. Under the condition of globally stagnating telecommunications markets, the challenge is to combine the two contrary objectives of investments in innovations with consistent cost management.

---

[6]The competition with cable providers works both-ways. Telecommunications operators are addressing customers of cable providers by offering IPTV services. Cable providers are addressing the customers of telecommunications operators by offering broadband internet services.

2.1.2   Emergence of Over-the-Top (OTT) Providers

The widespread adoption of mobile Internet access has lowered the barriers for many companies to enter the communication services market (Fritz et al. 2011, p. 269). Meanwhile, major Internet players have identified opportunities and have also entered these markets. In most cases these services are not necessarily expected to be major drivers of revenue growth; however, they are usually expected to complement the core business, similar to device sales or advertising. The most powerful Internet players are increasingly able to leverage their strengths in the value chain by presenting their communication services as the defaults in devices.

From a market perspective, OTT providers are the logical consequence of the changed market conditions. The rising emphasis of application services (Peppard and Rylander 2006, pp. 133–134) combined with the convergence in the ICT sector (Arlandis and Ciriani 2010, p. 121) have strengthened new competitors (Wulf and Zarnekow 2011a, pp. 290–292). From a technical perspective, the separation of application and communication services from their technical transportation (Knightson et al. 2005) has supported this trend. In practice, the impact of OTT providers on both telecommunications market and traditional telecommunications operators is discussed in various reports (cf. Table 2.1).

Telecommunications operators have several strategic options to overcome the challenges arising from OTT providers. Most of the strategies developed and implemented by telecommunications operators to deal with the pressure coming from OTT providers are defensive. The telecommunications operators are aware that OTT communication services are eroding their revenues and, therefore, they need to have a strategy in place to counteract this trend. Blocking VoIP services is a strategy that many telecommunications operators use.

Table 2.1 Selected reports about OTT market and strategies

Publisher	Title	Content	References
Analysys Mason	OTT communication services worldwide: stakeholder strategies	OTT trends and major players	Sale (2013)
Analysys Mason	Case study: Google's OTT communications strategy	Analysis of OTT services offered by Google	Bachelet and Sale (2014)
Informa Telecoms & Media	VoIP and IP messaging: Operator strategies to combat the threat from OTT providers	Evaluation of OTT markets for mobile service operators	Clark-Dickson and Talmesio (2013)
Strategy Analytics	Is VoLTE the answer to the OTT voice threat?	Impact of OTT VoIP services on mobile operator strategies	Kendall (2013)
IDATE Research	OTT video: Opportunities for Telcos around VoD, SVOD and Telco CDN	Analysis of market for OTT video services and impact on strategies of telecommunications operators	IDATE Research (2013)

Instead of blocking VoIP services, there are some mobile operators that are partnering with OTT providers, and also some mobile operators that are developing their own OTT-like services in their digital business divisions. So far, these two approaches represent the minority of cases. In particular, the attempt to develop own OTT-like services is a strategy which is still in its early stages and which will require a higher maturity level in the digital business areas. On the other hand, the current developments in the OTT market are increasing the pressure on telecommunications operators, giving them only a small window of opportunity to conceive an effective response strategy.

The strategic response alternatives for traditional telecommunications operators to OTT providers can be summarized as follows[7]:

- *Accept OTT services*: Several telecommunications operators have chosen a hands-off approach to any service that can increase the usage of data, including OTT services. These telecommunications operators believe that the non-occasional nature of communication services such as IP voice and messaging can lead to a strong incentive for customers to purchase a data plan upgrade.
- *Attack or absorb OTT services*: Many telecommunications operators have decided to attack OTT-based services directly by preventing subscribers from using IP services. This is realized by combining economic and technical aspects that prevent the use of IP services. Another approach is to absorb OTT services by making them ineffective from a customer's perspective. Customers use IP voice and messaging services with the objective to save money. In response, operators are, for example, introducing large voice and messaging bundles with the result that customers do not need to use OTT services in order to save money. In addition, offering services that are similar to the ones offered by OTT providers is a possible strategy. Launching proprietary OTT services is, so far, the least developed option. In the past decade, there have been some attempts by telecommunications operators to deploy instant messaging clients.
- *Partner with OTT providers*: In some cases, telecommunications operators decide to partner with OTT providers with the objective of benefiting from them. On the one hand, telecommunications operators are afraid that their core services could be marginalized by OTT providers; on the other hand, they are aware that these services can be popular amongst customers. Telecommunications operators that decide to partner with OTT providers might benefit from both the OTT services as well as the OTT brand.

The strategic options listed above are not necessarily mutually exclusive, and many telecommunications operators are active in several of these areas. Price will continue to be the major driver in the voice market. Therefore, telecommunications operators use pricing levers to ensure their voice services are relevant to most smartphone users.

---

[7]Based on results of Detecon's OTT knowledge development team. Please see also the reports listed in Table 2.1 for further details.

Google is an example of a successful OTT provider (Bachelet and Sale 2014). In some areas it is a strong competitor of established telecommunications operators. Google has established comprehensive product and service categories for devices, operating systems, applications and services, content and advertisement so as to service their customers from one source. This provides Google with a competitive advantage in comparison to telecommunications operators specialized in selected categories only. Offering the existing application service via own mobile network capacities (e.g., realized as a *Mobile Virtual Network Operator*) could be a strategic option that would fit to the ongoing convergence of the whole ICT sector. For traditional telecommunications operators, however, the demand for communication services is directly linked to the existence of attractive content and applications: for example, the growing demand for mobile data services is based on the ever-increasing range of mobile content and applications by, e.g., Google.

This one example highlights the complex interrelation between OTT services and telecommunications operators. The extensive communication services portfolio of OTT providers, their level of control and also the ability to monetize their services present a growing challenge for most telecommunications operators. There are still some operators that have not yet recognized the severe risk of their services being eroded by OTT-based communication services. However, the majority of telecommunications operators have clearly seen the urgent need for developing a strategy for OTT communications.

OTT's business models develop rapidly and change the traditional revenue models as follows[8]:

• Advertisement is one of the main revenue sources of OTTs;
• Paid subscriptions start to work for OTTs with a larger customer base;
• "Freemium" apps have proved to be an innovative monetization strategy;
• Cloud storage as an add-on service has increased profitability; and
• Business intelligence is a powerful tool for content distributors.

In Fig. 2.3 a phased approach is outlined to assess the impact of OTTs on the business and thus develop an effective, feasible response strategy tailored to the specific needs.

Several telecommunications operators are investing in the development of products and services for vertical markets like energy, automotive, healthcare, and education in order to generate additional revenue streams besides the traditional telecommunications business. In Sect. 2.1.3, the growth potential in vertical markets is further analyzed and concrete examples for selected vertical market service offerings are provided.

---

[8]Based on results of Detecon's OTT knowledge development team.

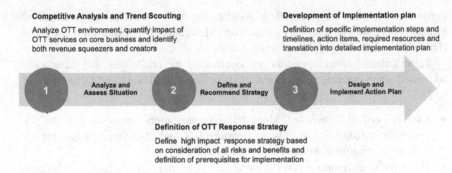

Fig. 2.3 OTT response strategy development approach

2.1.3 Growth Potential in Vertical Markets

In Sects. 2.1.1 and 2.1.2, the challenges facing telecommunications operators due to price decrease, cost pressure, and the threat posed by the emergence of OTT providers are explained. Telecommunications operators could address these challenges by generating new revenue streams in non-telecommunications business areas. Telecommunications operators have started to look into various vertical markets for which vertical-specific products and services can be developed and offered.

The common vertical markets named by most telecommunications operators are automotive, banking, consumer packaged goods, education, energy and utilities, government, healthcare, insurance, manufacturing, mining, public sector, retail, transportation and logistics as well as smart home. Cloud-based solutions and Machine-to-Machine (M2M) solutions are, for instance, services that can be offered to various verticals.

Several elements are required for telecommunications operators who decide to enter vertical markets including (Sapien 2011, p. 4):

- transformation capabilities beyond telecommunications;
- overview of product demands for vertical markets;
- innovative products and services to be offered;
- product development team with vertical knowledge;
- strong partner network for different verticals; and
- direct or indirect sales channels.

An analysis (Foong and Delcroix 2011) shows that services in vertical markets are expected to generate revenue amounting to 8.1 % of worldwide traditional telecom services revenue in 2015. Ambitious telecommunications operators are able to raise this figure up to 15 % or even 20 %. Media/entertainment (including advertising), Machine-to-Machine (M2M) services, cloud computing and IT services are promising areas for generating revenue (Foong and Delcroix 2011, p. 1). So far, most telecommunications operators are still facing several difficulties in running a profitable business in their vertical markets. On the cost side, major upfront investments are required. In most cases, on the revenue side telecommunications operators have to rely on indirect sales channels because their own sales channels need time to build up vertical sales capabilities.

Telecommunications operators are confronted by challenges that have to be addressed when entering new vertical markets including (Foong and Delcroix 2011, p. 7):

- *Lack of vertical knowledge.* Many telecommunications operators lack the necessary knowledge, know-how and capabilities and, therefore, partnerships as well as acquisitions should be considered.
- *Difficulty in developing vertical products.* A prerequisite to realize substantial revenues is the development of the right vertical products that actually meet the customer demands.
- *Presence of global competitors.* There are large, established global players with the required vertical expertise and customer base in various markets that are competing with the telecommunications operator.
- *Lack of global scale.* Regional telecommunications operators are less attractive to content and application developers. Content and application providers are more attracted by partners with global reach.

Partnerships, acquisitions, and strategic investments will play a significant role in this context and will also be a major driver for entering new vertical markets. In Sect. 2.3.2, the motivation for operator partnering, potential operator partnering areas and related benefits are described.

In general, a large number of initiatives in a vertical market do not necessarily correlate with a high maturity level of these initiatives. This effect is particularly the case in vertical markets that are exposed to strong influencing factors beyond control of the telecommunications operator (e.g., mobile health and mobile financial services). These verticals are indicating a greater need to tailor each product to specific market conditions. There is no single vertical market that has until today achieved the desired maturity. Figure 2.4 illustratively shows the correlation between the number of initiatives by vertical, and the average initiative readiness score (Velasco-Castillo and Renesse 2014, p. 12). Based on project experience with leading international telecommunications operators that is related to the establishment of M2M competence centers and cloud business units, there is an indication that these two initiatives will have the potential to reach the strategic target of achieving both a high number of initiatives and a high average readiness score.

Telecommunications operators are transitioning from a product-centric approach, in which all customers are offered the same service, to a customer-centric one. Customer-centric approach means designing customized solutions, tailored to the needs of each customer or customer-segment, which could be a specific vertical industry. Customized services will allow telecommunications operators to distinguish themselves and market unique solutions. This approach also changes the way telecommunications operators are organized, and they will

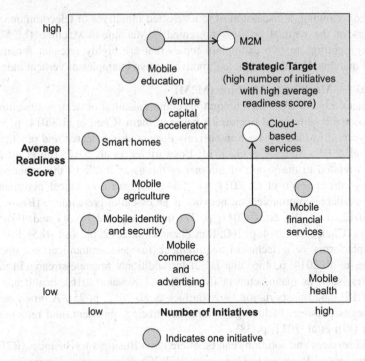

Fig. 2.4 Correlation of number of vertical initiatives and readiness score (according to Velasco-Castillo and Renesse 2014, p. 12)[9]

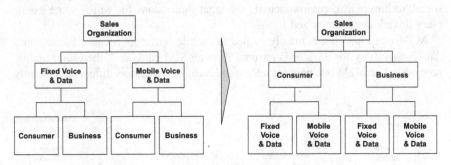

Fig. 2.5 Transition from product-centric to customer-centric organization (according to Pouillot 2013, p. 22)

typically be moving from a product-oriented sales structure to a customer-centric one (Pouillot 2013, p. 22) (cf. Fig. 2.5).

---

[9]In this figure the interrelation between the number of initiatives and the average readiness score is based on Velasco-Castillo and Rendesse (2014, p. 12). A strategic target window is added in the top right corner of the figure to highlight the strategic goal for all initiatives. Based on own project experience, M2M and cloud-based initiatives are highlighted to achieve the strategic goals in the first place.

In the following, a discussion of four selected initiatives of telecommunications operators in the vertical areas is presented.[10] Machine-to-Machine (M2M) and cloud computing are two exemplary topics that are highly relevant for entering vertical markets. Healthcare and automotive are two examples of vertical industries.

Vertical 1—Machine-to-Machine (M2M)

The general idea of M2M is a ubiquitous communication of devices (machines) in order to enable automated operations between them (Chen et al. 2014, p. 98). In recent years, this idea has been intensively discussed in research and practice (e.g. Ahn et al. 2010; Antunes et al. 2014; Boswarthick et al. 2012; Liu et al. 2015). M2M is related to the vision of *internet of things* as it allows the connection of everyday objects (Wu et al. 2011, pp. 36–37). From a technical perspective, a widely available communication network is an essential prerequisite (Boswarthick et al. 2012, p. 3; Wu et al. 2011, p. 37) that is enabled by 3G and 4G mobile networks (Chen et al. 2014, p. 100; Kan Zheng et al. 2012, pp. 184–185). Providing M2M platforms is a technical requirement for telecommunications operators (Antunes et al. 2014, p. 436) that facilitates additional revenue streams in various industries, such as manufacturing (Matsuda and Kosaka 2016), healthcare (Park et al. 2015), and transportation (Boswarthick et al. 2012, p. 25). A broad range of use cases is possible, including tracking and tracing, payment, and remote maintenance (Wu et al. 2011, p. 38).

M2M services and solutions can be offered in a Business-to-Business (B2B) and in a Business-to-Business-to-Consumer (B2B2C) environment. Figure 2.6 shows examples for B2B2C and B2B verticals that are relevant for M2M.

The main drivers for the M2M business are derived from political, economic, social, technological, environmental, and legal dimensions for which some exemplary drivers are summarized in Fig. 2.7.

M2M is a topic that is mainly related to mobile operators because the required data connectivity for devices is ensured through SIM cards and, therefore, mobile networks. The M2M business for telecommunications operators differs significantly

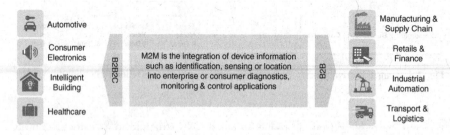

Fig. 2.6 M2M verticals for B2B2C and B2B

---

[10]Most of the information provided in this section is based on results from Detecon through project work in the international telecommunications industry.

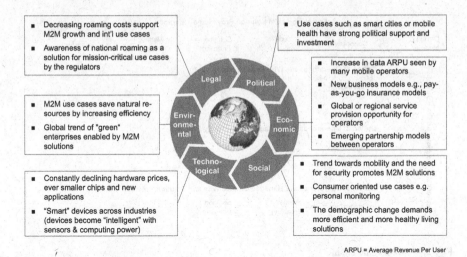

Fig. 2.7 Main drivers for M2M

	Traditional Mobile Operator	M2M Provider
Number of SIMs	Typically few SIM per customer	Many SIM per customer
Customer Interface	Manual, individual customer contact	Automated, with no customer contact on a per SIM basis
Bearer & Availability	Voice and data communications; medium requirements for availability	Data communication (IP/SMS) - barely voice; Significant use cases require high availability
Bandwidth /Volume	High bandwidth requirements and volume of data through mobile Internet	M2M use cases typically have low volume and bandwidth requirements per unit basis
Roaming	High domestic share, rather low international roaming requirements	Use cases with high international roaming volume; permanent roaming for availability
Value Chain Complexity	Typically E2E service offering by Operator	Operator is part of a fragmented M2M value chain; E2E orchestration skills required
Churn	Prepaid deals and short contract duration cause high turnover and high acquisition cost	Long contract duration (typ. 3-5 years) and high service provider switching costs

Fig. 2.8 Traditional Mobile Operator vs. M2M Provider

from the traditional business of a mobile operator. Figure 2.8 provides an overview of some differences for selected elements like the number of SIMs, customer interface, and roaming.

In the M2M business, telecommunications operators have the possibility to either become a *Wholesale Provider*, a *Managed Connectivity Provider* or a *Managed Connectivity and Solution Provider*. Those telecommunications operators providing M2M connectivity as a wholesale provider might only be able to realize small margins. Based on project experience in the M2M field, it is estimated that the highest margins can be realized through the provisioning of M2M solutions. Hence,

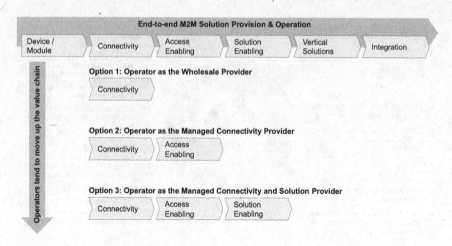

Fig. 2.9 M2M value chain

a strategic option for telecommunications operators is to move up the value chain from connectivity providers to access and solution enablers in order to capture higher margins. The classical M2M value chain for telecommunications operators from wholesale providers to managed connectivity and solution providers is presented in Fig. 2.9.

Vertical 2—Cloud Computing

Cloud computing is an intensively discussed research topic with a high practical impact (e.g. Leung et al. 2015; Trovati et al. 2015; Vijayakumar and Neelanarayanan 2016). It can be seen as an enabler that has tremendous impact on the whole ICT industry (Armbrust et al. 2010, p. 50). Computing services are decoupled from the technical capabilities (software and hardware) that are required to run these services (Buyya et al. 2009, p. 599). Today a huge amount of cloud services is available, delivered by providers through data centers hosting cloud applications that are accessed by customers via a network like, e.g., the Internet (Buyya et al. 2009, p. 600; Qian et al. 2009, p. 627).

This trend has tremendous implications for telecommunications operators from the technical and business perspective (Claus et al. 2010). First, cloud computing is enabled by the ubiquity of broadband telecommunications networks (Buyya et al. 2009, p. 600; Develder et al. 2012, p. 1151; Mikkilineni and Sarathy 2009, p. 57). Second, cloud computing offers virtualization capabilities that might influence the managing and provisioning of network services (Jain and Paul 2013, pp. 24–25). Third, the offering of cloud computing services is an opportunity for telecommunications operators (Claus et al. 2010, pp. 7–8). Hence, in recent years telecommunications operators have significantly invested in the area of cloud computing in order to be able to provide cloud services to consumers and business customers across various verticals (Claus et al. 2010). Furthermore, cloud computing can also be considered as an enabler for growth in vertical markets and is related to other vertical topics such as M2M (Chen et al. 2014, p. 104; Wu et al. 2011, p. 37).

Various guidelines and models for the development and deployment of cloud-related services already exist. In the telecommunications industry the definitions, service models, and deployment models provided by the *National Institute of Standards and Technology* (*NIST*) are widely used. According to the NIST, the typical cloud service models are (Mell and Grance 2011, pp. 2–3):

- *Infrastructure as a Service* (*IaaS*) provides fundamental computing resources (e.g., storage) that can be used to run any software.
- *Platform as a Service* (*PaaS*) offers the deployment of user-created applications over the cloud infrastructure, for example, by using programming libraries or tools.
- *Software as a Service* (*SaaS*) covers the whole deployment of applications that can be accessed by users.

According to the NIST, those services should include on-demand self-service, possibilities for access through heterogeneous clients, and scalability mechanisms. Based on the target group, the deployment could be realized as a private, community, or public cloud. Also a combination of those approaches is possible (Mell and Grance 2011, pp. 2–3).

Vertical 2—Healthcare
The healthcare market is complex and is characterized through a high number of stakeholders that are part of the value chain. The key stakeholders in the healthcare ecosystem include hospitals, general practitioners, health insurances, pharmaceutical companies, health ministries, providers of ICT solutions (i.e., hardware and software), and the patients themselves. The healthcare market distinguishes between the primary market, the secondary market, the tertiary market and new markets as shown in Fig. 2.10.

The current transformation of the healthcare market leads to significant investments along the value chain. In this context, ICT is an enabler for automation, data security and privacy, integration of different standards, telemedicine, patient self-monitoring platforms, digital health insurance cards and health commerce. As an example, according to the Gulf Cooperation Council (GCC), the healthcare spending in the Middle East is expected to increase fivefold, from US\$12 billion in 2007 to more than US\$60 billion by 2025, and ICT is expected to be fastest growing in the healthcare area (Mourshed et al. 2014).

Telecommunications operators are well known for their capabilities to develop, implement, and integrate different ICT solutions as well as to handle huge amounts of customer data. Hence, some telecommunications operators have started to invest in the development, implementation, and market launch of ICT solutions for the healthcare market. The vertical market for healthcare is an opportunity for the telecommunications operators to partly escape from the revenue decline and cost pressure described in Sect. 2.1.1.

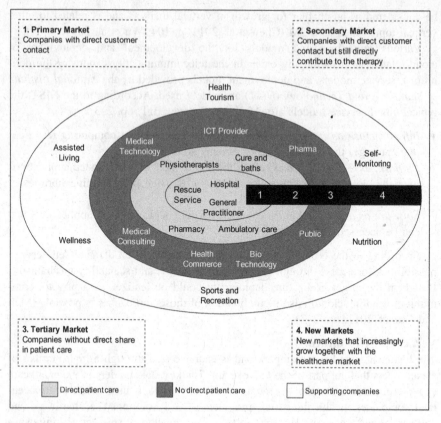

Fig. 2.10 Market map for healthcare

As an example, Deutsche Telekom, Germany's largest fixed and mobile operator, has done significant investments to successfully enter the healthcare market. Since 2010 the healthcare division has grown rapidly at Deutsche Telekom. It has developed many individual eHealth products on its own, recruited new partners, invested in start-up companies, and made successful acquisitions. They have done the investments on such a large scale that Deutsche Telekom Healthcare & Security Solutions GmbH is now one of Europe's healthcare ICT market leaders. Healthcare product areas covered include, e.g., connected healthcare, telemedicine, diabetes prevention portals, hospital information systems (HIS), adherence solutions, patient entertainment and digital insurance cards.[11]

Vertical 4—Automotive
Today the car is an essential part of people's connected life and work. With state-of-the-art ICT, driving becomes more efficient, safer, and more convenient.

---

[11]Please see www.telekom-healthcare.com for further details.

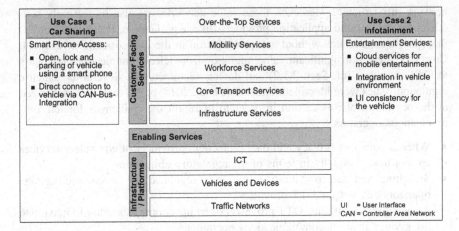

Fig. 2.11 Overview of connected mobility services in automotive

Consequently, the automotive industry offers significant business potential for telecommunications operators. As leading ICT service providers with their entire mobility ecosystem, high-performance mobile communications network, high standards of security, and quality, telecommunications operators could be the perfect partner for the automotive industry.

The advantage for automotive manufacturers is that they gain permanent, direct access to their customers and can manage services online by accessing their vehicles remotely. Once the car is part of a mobile network, automotive manufacturers can save distribution and service costs, continuously improve their product quality, tie customers into their car workshop, and control and improve their capacity utilization. Also, fleet operators benefit by integrating vehicle-based business processes that support more efficient, more sustainable use of their vehicles, and logistics service providers can optimize the operating costs of their trucks, route planning and real-time truckload management if their vehicles are online. Connected mobility services provided by telecommunications operators are based on the interworking of ICT and infrastructure components. As depicted in Fig. 2.11, use cases for connected mobility services include car sharing, infotainment and connected mobility experience.

2.1.4 A New Role for Regulators

The uneven playing field in the digital services ecosystem hinders network owners from capturing fair returns, or even the returns they had expected. The increasing demand for high transmission bandwidths requires extensive investments in network infrastructure. However, those same networks are also increasingly beneficial for content and application providers (like, e.g., Google, Facebook, Netflix) that gain high revenues without any participation in the infrastructure investment. Currently, there is a significant value migration from telecommunications operators to OTT providers (cf. Sect. 2.1.2) and device manufacturers. Several asymmetric regulation issues were identified (Amendola et al. 2014, p. 22) that are summarized at the beginning of this section in order to highlight the challenges for telecommunications operators:

- When it comes to privacy and data protection, providers of equivalent services are not treated equally in terms of the regulatory obligations.
- Switching and data portability is currently regulated for telecommunications operators and not for OTT providers.
- As new market entrants, OTT providers often have more flexibility to maximize tax savings than telecommunications operators.
- Some services provided by OTTs are not subject to the strict e-communications services rules.

The amount of data traffic generated over mobile networks by applications is constantly increasing. The impact of the growth of mobile smart devices and connections on global traffic has been analyzed by Cisco (2015). Traffic of smart devices is expected to grow from 88 % of the total global mobile traffic to 97 % by 2019. This percentage is significantly higher than the ratio of smart devices and connections, which is estimated to reach 54 % by 2019 (Cisco 2015, p. 10). The main reason for this is that, on average, a smart device generates much more traffic than a non-smart device. As a consequence, backbone networks are required to handle the explosion in data (e.g., through fiber optic technologies).

Another central trend has been explored regarding social media and social networking (ITU 2012, p. 5). The number of active social media users surpassed the first billion already in 2011, and most of them connect to social media using their mobile devices. An interesting finding is that the countries with the ten highest penetrations of social media users are located in developing countries. The profile of users is also changing, with a growing number of organizations, public entities, telecom/ICT regulators, and government agencies joining the individual and business users (ITU 2012, p. 5). It is a fact that social media has emerged in recent years as a tool for hundreds of millions of Internet users worldwide. Regulators should consider social media from several perspectives. The social media usage must be better understood by regulators so that the importance of social media can be properly assessed for policy development purposes (ITU 2012, p. 14). For regulators, it is also important to assess whether social media raises new regulatory or policy challenges that have to be addressed. It is relatively certain that regulators will be required to establish a policy framework for the use of social media in the near future.

Regulation	Licensed network operator	OTT player
Licensing	Subject to license and license fee	No license required
Quality of Service	SLAs included in the license	No quality requirements
Interconnection	Interconnection mandated	No interconnect requirements
Universal Service	Subject to universal service obligation	Not subject to universal service regime
Consumer Protection	Subject to consumer protection policy	Little or no enforcement power
Legal Interception	Usually license condition	Country dependent
Taxation	Subject to national tax regime	Service dependent

Fig. 2.12 Regulatory imbalance for operators and OTTs

On the one hand, there is an apparent imbalance regarding market and market entry conditions between licensed telecommunications operators and OTT providers. On the other hand, a new regulatory balance is not yet in sight. A comparison[12] of the regulatory obligations for licensed network operators and OTT providers has been performed, and the results of the comparison are summarized in Fig. 2.12.

In this context, the concept of *net neutrality* is further examined. Net neutrality is a somewhat vague concept. A common and at the same time high-level understanding of net neutrality is that all IP traffic should be treated equally, regardless of the type of content, service, application or device. There is an intense discussion about the concept of net neutrality (e.g. Belli and De Filippi 2015; Plunkett 2014, p. 10). ITU has defined network neutrality as follows: "Network neutrality is best defined as a network design principle. The idea is that a maximally useful public information network aspires to treat all content, sites, and platforms equally. This allows the network to carry every form of information and support every kind of application." (ITU 2013, p. 16).

---

[12]The comparison is based on project work conducted by Detecon.

Open and unrestricted access to the Internet

- Non-discriminatory access to any services and to all content available on the Internet
- Openness as social and policy goal
- Openness as innovation facilitator

Introduction of Internet service classes

- New business models to recover infrastructure investment and operations
- Introduction of QoS parameters such as "best effort", "critical", and "real time"
- Reversion current network deployment approach based on over-provisioning

Fig. 2.13 Network neutrality and two opposing positions[13]

There are a number of issues related to network neutrality which regulatory authorities should consider. The focus to date has been at the national level. However, the Internet is in fact a global network. It seems inevitable that, at some point, there will be a push to extend the regulation of net neutrality from the national to the international level (ITU 2013, p. 22). The following three different categories of actions can be differentiated:

- *Cautious observation*: These countries have considered whether network neutrality rules are needed at this point in time and decided not to take any action for now.
- *Tentative refinement*: These countries have implemented a "light" approach that introduces some new rules to the existing regulatory framework governing communications services. For example, some rules require greater transparency and disclosure of network management practices. Still, these new rules do not go so far as to prohibit certain behaviors.
- *Active reform*: These countries have gone further with the changes to their regulatory framework and prohibit specific behaviors by Internet Service Providers (ISPs). The changes to the regulatory framework include, for example, the prohibition of blocking and throttling that are often subject to reasonable network management practices.

The network neutrality debate can be characterized by two opposing positions. One position is open and unrestricted access to the Internet, whereas the other position is about the introduction of Internet service classes. It is recommended that regulators should go beyond the current either-or-approach, but combine the two positions as outlined in Fig. 2.13.

---

[13]Network neutrality and two opposing positions based on project work from Detecon.

2.2 Telecommunications Products and Services

In order to react to the changed market conditions, telecommunications operators are confronted with continuous innovations (Picot 2006) and shorter product development cycles (Bruce et al. 2008). Realizing a flexible interrelation between commercial products and technical services is important for a fast reaction to market demands. Furthermore, due to deregulated markets and increased competition, committed focus on the customer is essential. In Sect. 2.2.1, a general product and service structure is introduced. The consistent management of the customer experience is discussed in Sect. 2.2.2.

2.2.1 Interrelation Between Commercial Products and Technical Services[14]

Telecommunication services are services that are usually provided fully or predominantly in telecommunication networks. In a broader sense services are, in the context of distributed systems, described as a component that provides certain functionality to a user (Coulouris et al. 2005, pp. 7–8). A hierarchical decomposition of services is possible because communication systems are represented in different layers (e.g., Open Systems Interconnection, OSI). This means that the service of a communication system is composed of services from individual layers as shown in Fig. 2.14 (Tanenbaum and Wetherall 2014, p. 30).

Services should therefore be described with reference to a consistent level of detail. A service of the transmission layer has to be distinguished from a service of the application layer, although both services might contribute to the same communication service. In this context protocols are understood as a framework of rules to execute a particular service (Tanenbaum and Wetherall 2014, pp. 40–41). At the same time, protocols within a layer can be arbitrarily changed as long as the service is executed towards the user according to the agreed quality parameters. Exemplary protocols are the well-known *Transmission Control Protocol/Internet Protocol* (TCP/IP) used in the Internet (Tanenbaum and Wetherall 2014, p. 41) and the *Radio Link Control/Medium Access Control* (RLC/MAC) used in mobile telephony (Werner 2010, p. 193). From a technical perspective, a telecommunication service is a communication facility that is described through distinct features (e.g., information type, communication type, bandwidth requirement) and service performance.

Previously, dedicated service networks were operated, meaning that networks were assigned to one dedicated service—e.g., the telephone network was assigned to the telephony service. This arrangement was no longer necessary with the usage of digital networks, which is also referred to as technology convergence

---

[14]This section is a translated and slightly revised version of the content published in Czarnecki (2013, pp. 39–41).

Fig. 2.14 Relation between
services and layers (according
to Georg 1996, p. 43)[15]

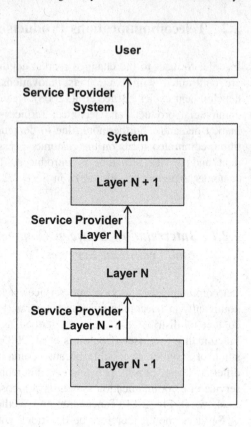

(Wieland 2007, p. 46). However, this has also made the classification of
telecommunication services difficult, as they are strongly dependent on the technical development and the usage behavior. For instance, the differentiation between
mobile telephony and data services was quite useful at the end of the 20th century
(Gerpott 1999, p. 61). Nowadays, with the increasing usage of mobile data services,
this differentiation is no longer sustainable.

With respect to communication types, a differentiation can be made between
individual services (e.g., voice telephony) and a distribution service (e.g., radio).
The individual service facilitates an information exchange between two or more
participants in both directions. The distribution service allows asymmetric information exchange from one sender to several recipients (Gerpott 1999, pp. 59–60).
In addition, it can be distinguished by the form of exchanged information (e.g.,
voice, picture, text).

The selling of telecommunication services to customers remains an original
objective of a telecommunications operator. In this respect, the consideration of
telecommunication services from a marketing perspective also has to be made.

---

[15]Translated version of the illustration published in Czarnecki (2013, p. 39).

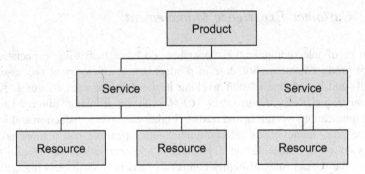

Fig. 2.15 Interrelation between product, service, and resource (according to Bruce et al. 2008, p. 19; Czarnecki and Spiliopoulou 2012, p. 393; TM Forum 2015, p. 46)[16]

At the same time, a differentiation between a general service (e.g., the installation service of a service technician), a technical service (e.g., the transmission service of a mobile network) and a service as a product (e.g., consisting of general and technical services) is important. In fact, the term *service* is used for all three different types. The differentiation is, in most cases, only possible by considering the overall context. In the technical context of IT systems or communication networks, services are often understood as the technical provision of functionality as described above. From an economic perspective, the telecommunications industry belongs to the service industry and, accordingly, a service will be provided to the customer (Zeithaml and Bitner 2003, p. 3). As shown in Fig. 2.14, only a subset of the services is perceived by the customer. All other services are executed within the telecommunications network as well as the telecommunications system, and are not visible to the customer.

In order to avoid terminological confusion, the term *product* should be used to describe a telecommunication service that is provided to a customer (Bruce et al. 2008, p. 19; Snoeck and Michiels 2002, p. 335). According to Bruce et al. (2008, p. 19) and TM Forum (2015, p. 46), the following tripartition is used for structuring (cf. Fig. 2.15):

- *Product* represents the commercial view and can consist of one or several services (in a broader sense) and technical devices (e.g., telephone)
- *Service* (in a broader sense) is a detailing of a product and can comprise a technical telecommunication service (e.g., voice telephony) as well as an additional provision of service (e.g., connection of telephone)
- *Resource* represents the lowest level of detail and, therefore, the building blocks of services. A resource can be a physical device that is owned by the customer (e.g., telephone) or it can be used by the customer either completely or partially (e.g., telephone line). Resources can also be immaterial goods (e.g., installation work).

---

[16]Translated version of the illustration published in Czarnecki (2013, p. 41).

2.2.2 Customer Experience Management

Customers of telecommunications operators do have increasing expectations in terms of product functionality, ease of product usage, efficiency of processes, and the skills and knowledge of staff working in the different sales channels. A solid *Customer Experience Management* (*CEM*) can be a major differentiator for telecommunications operators and leads to higher customer satisfaction and loyalty.

In the past, monopolistic telecommunications operators had a more administrative view on customer demands. Nowadays telecommunications operators have to accept the typical rules of highly competitive markets: customers are willing to pay more if a good service quality is ensured, whereas weak service experience leads to complaints and customer churn. Some reasons for customer churn can be the interaction with unmotivated employees, unexpected charges, or products and services with a poor quality. It is important that telecommunications operators understand that negative customer experience pushes customers away. At the same time, it is human nature to pass on negative experience more intensively to others than positive experience. Hence, a primary objective of telecommunications operators should be to introduce enhanced processes and solutions which will minimize the probability of negative customer service.

High customer satisfaction and loyalty at all customer touch points is the key objective of any CEM endeavor. A comprehensive approach is needed to achieve sustainable optimization of customer interactions and avoid customer disappointments in critical interactions—the so-called *moments of truth*. This new, overall approach should aim for a fundamental change in the customers' perspective and the avoidance of customer disasters in future by concentrating on critical interaction points. The change in the customers' perspective will result in the sustainable optimization of critical customer interactions. Concentrating on high priority customer grievances, rather than overambitious and complex CEM concepts that often fail during the implementation, usually leads to tangible and measurable benefits in a relatively short timeframe. Practical project experience has shown that the trial and validation of pilots can be more successful than doing endless analysis. A general rule of this approach is that there should always be careful consideration of the customer perspective before undertaking any optimization measures.

The approach developed and successfully applied to improve customer experience and hence customer satisfaction consists of two steps:

In the first step, key pain points with high priority from a customer perspective are identified at different customer interaction points. The customer interaction points can be identified by analyzing customer-centric processes, such as order, change, termination, or problem solving (cf. Sect. 4.3.1). Figure 2.16 illustratively shows some customer-centric processes, typical customer interaction points, and examples for the prioritization of interaction points from a customer perspective. Customer interaction points with a high priority and a negative customer perception are selected to be addressed first.

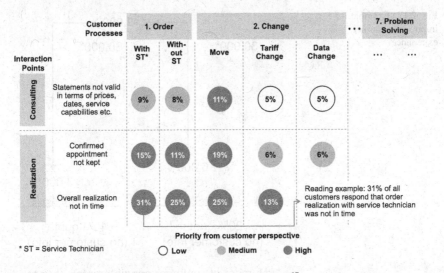

Fig. 2.16 Pain point identification from a customer perspective[17]

In the second step, action areas and concrete initiatives for eliminating negative moments of truth are defined to implement optimization measures with tangible benefits. For each initiative, a problem statement has to be formulated in order to ensure that all stakeholders who are involved in the initiative have the same understanding of the existing challenges. The specific objectives and the expected outcome of each initiative also have to be defined amongst the stakeholders. All initiatives should strictly focus on customer anger elimination.

Next, there is an illustrative example related to the problem-solving process of a telecommunications operator. This example consists of a problem statement, the formulation of the initiative objective, and the expected benefits:

Problem Statement: 20 % of all customers are not proactively informed when a service technician is unable to keep an appointment.
Objective: To reduce the non-information quota by 50 % in cases involving delay. Customer grievances can be significantly reduced if adequate information is sent to customers in advance in situations where a service technician cannot keep an appointment.
Benefits (illustrative figures): Fig. 2.17 shows that a 50 % reduction in the non-information quota in those situations involving any delay would lead to 3000 additional customers getting a message when a technician is unable to keep an appointment.

---

[17]Approach developed by Detecon in cooperation with a leading European telecommunications operator.

Fig. 2.17 Example for improved customer experience

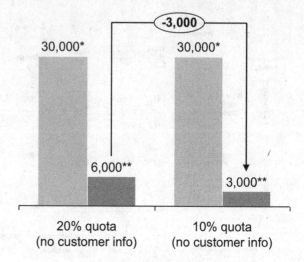

* Number of customer appointments per year
** Number of customers not informed when an appointment will not be kept

Greater benefits can be achieved if similar measures are defined and implemented for many, different high priority customer grievances. Measures could include the proactive disposition of customer appointments in the technical service area, or the usage of existing IT-systems and relevant customer data to contact the customer. In practice, this is unfortunately often not the case in today's telecommunications industry. A positive effect of such measures is that any delay and rescheduling of customer appointment become transparent to the customer as soon as possible. This subsequently leads to the improvement of perceived appointment compliance and a reduction in the general dissatisfaction with rescheduling.

2.3 Telecommunications Value Chain

The value creation in the telecommunications industry is heavily influenced by new players, such as content and applications providers (Peppard and Rylander 2006, pp. 128–129; Pousttchi and Hufenbach 2011, p. 307). The resulting erosion of the traditional value chain is discussed in Sect. 2.3.1. A reaction of telecommunications operators is the establishment of new partnerships (Grover and Saeed 2003, pp. 121–125). The impact and a step-wise partnering approach are introduced in Sect. 2.3.2.

2.3.1 Erosion of the Traditional Telecommunications Value Chain

For a long time, the focus of the telecommunications industry has been on the transmission of information over long-distance networks. The transmission was mainly focused on the technically correct transmission of signals. A major part of the value creation of a telecommunications operator was the roll-out and operations of the required network infrastructure. Those communications networks were related to extensive long-term investments, which served as an enormous market entry barrier for new competitors. The requisite skillset was mainly related to communications engineering.

However, in the last two decades the telecommunications industry has gone through a major transformation (Cave et al. 2002, p. 3). The driver of this transformation is the technological development in terms of higher bandwidth and improved computing power. This technological development has resulted in innovations and a different user behavior—for example, through social networks (Picot 2007, p. 19). The convergence of telecommunication services (Bertin and Crespi 2009, pp. 188–189), the usage of mobile value-added services (Bina and Giaglis 2007, pp. 241–246), and the impact of mobile devices (i.e., smartphones) with high performance operating systems (Basole and Karla 2011) are all impacting the telecommunications industry. The value creation in telecommunications has moved away from the pure transmission of information towards the offering of application services (Peppard and Rylander 2006, pp. 133–134; Pousttchi and Hufenbach 2011, p. 299). The consolidation of telecommunication, computer and media industry is a result (Arlandis and Ciriani 2010, p. 121).

The telecommunications value chain creates exciting new opportunities and new challenges for infrastructure and service providers at the same time. The value chain that has long been successfully established in the telecommunications industry for a long time is increasingly being deconstructed. New, powerful players are entering the market, and a radical restructuring of the industry is ongoing. In fact, the rapid technological developments and increasing market turbulences have added new dimensions to an already complex scenario. Several implemented business models that were generating revenues for telecommunications operators have become less important (Li and Whalley 2002, p. 460). The increased focus on applications has resulted in a convergence of voice, video, and data. The technical transmission becomes a minor part of the overall telecommunications value chain, which is now confronted with new players, mergers, and acquisitions (Tardiff 2007, p. 132; Wulf and Zarnekow 2011b, pp. 10–11). Entertainment services like TV offers are linked to traditional communication services, leading to new competition between TV cable operators and communication network operators (Plunkett 2014, p. 7). Virtual

Table 2.2 Selected reports about changes of the telecommunications value chain

Publisher	Title	Content	References
STL Partners	Five principles for disruptive strategy	Strategic options for business models in the telecommunications industry	STL Partners (2014)
Ovum	Innovative broadband pricing strategies	Importance of content and applications for differentiated pricing strategies	Ovum (2014a)
Ovum	Digital operator strategies	Evaluation of business models based monetization of new services	Ovum (2014b)
Informa Telecoms & Media	Industry outlook 2014—digital futures: Creating new roles and value chains	Broad analysis of changed market conditions, e.g., spend per different players of the value chain	Informa Telecoms & Media (2014)
IDATE	Future telecom: Trends and scenarios for 2025	Evaluation of future scenarios of the value chain and their impact on telecommunications operators	IDATE (2014)

business models (e.g. Virtual Mobile Operator) exist that allow a successful value creation without owning and operating a communication network (Pousttchi and Hufenbach 2009, p. 87). Li and Whalley (2002, pp. 462–468) argue that the result is a value network consisting of software intermediaries, financial intermediaries, content providers, portals, and resellers.

Also, in practice, the changes of the telecommunications value chain and the impact of those changes on business models are discussed in various reports (cf. Table 2.2). All of those reports describe the erosion of the value chain and the requirements of new, changed business models for traditional telecommunications operators. The pure provisioning of voice and data transmission via fixed or mobile networks seems to be an outdated business model. The change from usage-dependent to flat-rate tariffs was the starting point for traditional telecommunications operators to think about new revenue streams. Various studies illustrate those changed market conditions based on revenue and usage figures (e.g. Plunkett 2014). As a result, applications become an important part of the value creation (cf. Fig. 2.18). The combination of transmission services with application services allows differentiated pricing strategies and new revenue models (e.g., advertisement). For traditional telecommunications operators, such innovations require investments into own developments, acquisitions, or partnerships with new market players (cf. Sect. 2.3.2).

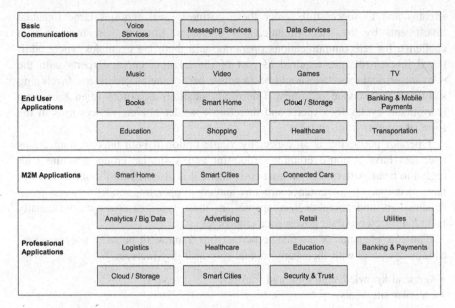

Fig. 2.18 Selected innovative services of the telecommunications value creation[18]

2.3.2 The Operator Partnering Imperative

The way towards sustainable growth for telecommunications operators in existing and new business areas still remains a major challenge. Nowadays telecommunications operators are facing a twofold competition, from well-established telecommunications operators and from large and small players in the market, who successfully attract consumers with mobile Internet and innovative online services, for example.

Through increasing competition in the telecommunications industry as well as the emergence of OTT providers as described in Sect. 2.1.2, the need for establishing partnerships between telecommunications operators is becoming more important. In the past, telecommunications operators have mainly concentrated on moving their own business forward without taking partnerships with other operators seriously into consideration. The situation has changed and leading telecommunications operators are becoming more open to establish strategic partnerships for high priority business areas.[19]

In Sect. 2.1.3, the general growth potential in vertical markets and the activities of telecommunications operators in selected business areas like M2M, healthcare, cloud, and automotive are discussed. The endeavor to generate new revenue

---

[18]Own illustration based on the reports shown in Table 2.2.

[19]The information provided in the section is mainly based on project work conducted by Detecon.

streams and to successfully enter these business areas requires large financial investments by the telecommunications operators. Entering new markets is a challenge for telecommunications operators, and there are examples where additional investments are required as part of the learning curve. Experts with the relevant vertical expertise should be an active part of the project teams developing solutions and service offerings. Telecommunications operators often face long recruitment cycles for experts and therefore consider alternative scenarios in the form of support from partners.

Operator partnering is an especially viable option if both telecommunications operators have a non-overlapping footprint and a similar group structure with regard to headquarters and national companies. It is also helpful if both companies have a comparable governance structure and equal operating model between group headquarters and operational companies, as these types of operators are usually facing similar corporate governance challenges.

The core elements of a strategic partnership framework between telecommunications operators with increasing mutual responsibility include:

- general knowledge sharing and transfer;
- regular site visits;
- joint business models;
- joint market and sales approach for products/services; and
- revenue and investment sharing.

Selected advantages for the telecommunications operators are:

- benefit from existing solutions of the other operator;
- joint product development and innovation activities;
- reduction of product development cost;
- new and innovative business models;
- potential to enter new regional markets; and
- footprint extension for own products and services.

For the purpose of identifying potential business areas for partnership, knowledge sharing, and transfer as well as the development of joint business models, it is recommended that telecommunications operators establish a strategic partnership framework for the mutual benefit of both organizations. While establishing the strategic partnership framework, investments, benefits, and the level of interest have to be in balance for both parties. A strategic partnership framework can be established by following a stepwise approach as shown in Fig. 2.19.

Step 1—Framework Agreement
In the first step, the telecommunications operators explore opportunities for commercial partnerships with regard to several business areas in the telecommunications market or other vertical markets, which should enable them to exploit untapped synergy effects. For this initial step, both telecommunications operators enter a general framework agreement. In the framework agreement, the operators agree to cooperate with each other and use reasonable commercial efforts to identify and evaluate possibilities for commercial partnerships in the areas of cooperation.

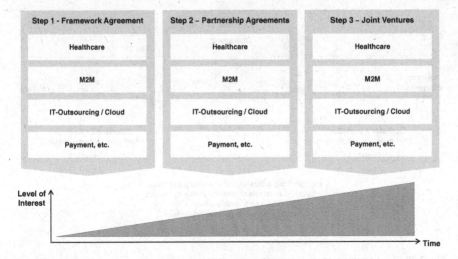

Fig. 2.19 Strategic partnership framework establishment

Step 2—Partnership Agreements

In the second step, concrete partnership agreements for joint business models are established between both telecommunications operators with the objective of revenue and investment sharing. The partnership agreements are ideal for cooperation in selected business areas that promise a positive return on investment. Besides the elaboration of joint business models and solutions, both telecommunications operators can also develop joint go-to-market and sales approaches. Through joint go-to-market and sales approaches, knowledge can be exchanged between both telecommunications operators. Lessons learned from the actual implementation of the market approaches for products and services can also be shared, and these are an important input for improvement initiatives in both telecommunications operators.

Step 3—Joint Ventures

In the third step, both telecommunications operators decide to establish a joint venture as a separate entity that will fully focus on the common business interest.

A real-life example for operator partnering and the establishment of a joint venture is BuyIn[20] which is the 50:50 procurement joint venture between Deutsche Telekom and Orange. The joint venture combines approximately EUR 20 billion of annual spend of the two companies in three main domains: network, customer equipment and service platforms. By pooling their procurement activities in this equal joint venture, Deutsche Telekom and Orange expect to achieve significant economies of scale and deliver annual savings through best price alignment, the aggregation of volumes, the harmonization of specifications and improved collaboration.

---

[20]Please see http://www.buyin.pro for further details.

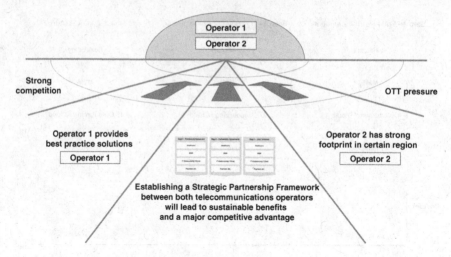

Fig. 2.20 Target picture of strategic partnership framework

As illustrated in Fig. 2.20, the strategic partnership framework consisting of the three steps will bring both telecommunications operators together in selected business areas. The target picture—i.e., to deal with competition from other operators and OTT pressure, as well as to jointly bring best practice solutions to the market—can all be achieved through implementing the strategic partnership framework by both telecommunications operators.

References

Ahn, J. Y., Song, J., Hwang, D.-J., & Kim, S. (2010). Trends in M2M application services based on a smart phone. In T. Kim, H.-K. Kim, M. K. Khan, A. Kiumi, W. Fang, & D. Ślęzak (Eds.), *Advances in software engineering* (pp. 50–56). Berlin Heidelberg: Springer.

Amendola, G., Gassot, Y., Lebourges, M., & Stumpf, U. (2014). *Re-thinking the EU telecom regulation.* Idate.

Antunes, M., Barraca, J. P., Gomes, D., Oliveira, P., & Aguiar, R. L. (2014). Unified platform for M2M telco providers. In R. Hervás, S. Lee, C. Nugent, & J. Bravo (Eds.), *Ubiquitous computing and ambient intelligence. Personalisation and user adapted services* (pp. 436–443). Cham: Springer International Publishing.

Arlandis, A., & Ciriani, S. (2010). How firms interact and perform in the ICT ecosystem? *Communications and Strategies*, 121–141.

Armbrust, M., Stoica, I., Zaharia, M., Fox, A., Griffith, R., Joseph, A. D., et al. (2010). A view of cloud computing. *Communications of the ACM, 53*, 50. doi:10.1145/1721654.1721672

Bachelet, C., & Sale, S. (2014). *Case study: Google's OTT communications strategy.*

Basole, R. C., & Karla, J. (2011). On the evolution of mobile platform ecosystem structure and strategy. *Business & Information Systems Engineering, 3*, 313–322. doi:10.1007/s12599-011-0174-4

Belli, L., & De Filippi, P. (2015). *The net neutrality compendium.* New York, NY: Springer Science+Business Media.

Bertin, E., & Crespi, N. (2009). Service business processes for the next generation of services: A required step to achieve service convergence. *Annals of Telecommunications, 64*, 187–196.

Bina, M., & Giaglis, G. (2007). Perceived value and usage patterns of mobile data services: A cross-cultural study. *Electronic Markets, 17*, 241–252. doi:10.1080/10196780701635773

Boswarthick, D., Elloumi, O., & Hersent, O. (2012). *M2M communications a systems approach.* Hoboken, N.J: ETSI, Wiley.

Bruce, G., Naughton, B., Trew, D., Parsons, M., & Robson, P. (2008). Streamlining the telco production line. *Journal of Telecommunications Management, 1*, 15–32.

Buyya, R., Yeo, C. S., Venugopal, S., Broberg, J., & Brandic, I. (2009). Cloud computing and emerging IT platforms: Vision, hype, and reality for delivering computing as the 5th utility. *Future Generation computer systems, 25*, 599–616. doi:10.1016/j.future.2008.12.001

Cave, M. E., Majumdar, S. K., & Vogelsang, I. (2002). Structure, regulation and competition in the telecommunication industry. In M. E. Cave, S. K. Majumdar, & I. Vogelsang (Eds.), *Structure, Regulation and Competition: Vol. 1. Handbook of telecommunications economics* (pp. 1–40). Amsterdam: Elsevier.

Chen, M., Wan, J., Gonzalez, S., Liao, X., & Leung, V. C. M. (2014). A survey of recent developments in home M2M networks. *IEEE Communications Surveys & Tutorials, 16*, 98–114. doi:10.1109/SURV.2013.110113.00249

Cisco. (2015). *Cisco visual networking index: Global mobile data traffic forecast update, 2014–2019.* San Francisco: Cisco.

Clark-Dickson, P., & Talmesio, D. (2013). *VoIP and IP messaging: Operator strategies to combat the threat from OTT players.*

Claus, T., Kellmereit, D., & Narielvala, Y. (2010). *The future of cloud: A roadmap of technology, product, and service innovations for telecoms.* San Francisco, CA: Thorsten Claus.

Copeland, R. (2009). *Converging NGN wireline and mobile 3G networks with IMS.* Boca Raton: CRC Press.

Coulouris, G. F., Dollimore, J., & Kindberg, T. (2005). *International Computer Science series: 4th ed. Distributed systems: Concepts and design.* Harlow, England; New York: Addison-Wesley.

Czarnecki, C. (2013). *Entwicklung einer referenzmodellbasierten Unternehmensarchitektur für die Telekommunikationsindustrie.* Berlin: Logos-Verl.

Czarnecki, C., & Spiliopoulou, M. (2012). A holistic framework for the implementation of a next generation network. *International Journal of Business Information Systems, 9*, 385–401.

Czarnecki, C., Winkelmann, A., & Spiliopoulou, M. (2013). Reference process flows for telecommunication companies: An extension of the eTOM model. *Business & Information Systems Engineering, 5*, 83–96. doi:10.1007/s12599-013-0250-z

Develder, C., De Leenheer, M., Dhoedt, B., Pickavet, M., Colle, D., De Turck, F., et al. (2012). Optical networks for grid and cloud computing applications. *Proceedings of the IEEE, 100*, 1149–1167. doi:10.1109/JPROC.2011.2179629

Doeblin, S., & Dowling, M. (2007). Horizontal und vertikal integrierte Geschäftsmodelle von Telekommunikationsanbietern und Service Providern. In A. Picot & A. Freyberg (Eds.), *Infrastruktur Und Services—Das Ende Einer Verbindung?* (pp. 29–41). Berlin, Heidelberg: Springer.

Ehrmann, T. (1999). Mark-und Wertschöpfungsstrukturen in der Telekommunikation. In D. Fink & A. Wilfert (Eds.), *Handbuch Telekommunikation Und Wirtschaft: Volkswirtschaftliche Und Betriebswirtschaftliche Perspektiven* (pp. 33–48). München: Verlag Franz Vahlen.

Foong, K. -Y., & Delcroix, J. -C. (2011). *Market trends: New revenue opportunities for telecom carriers in 2015.* Stamford: Gartner.

Fransman, M. (2002). Mapping the evolving telecoms industry: The uses and shortcomings of the layer model. *Telecommunications Policy, 26*, 473–483. doi:10.1016/S0308-5961(02)00027-7

Fritz, M., Schlereth, C., & Figge, S. (2011). Empirical evaluation of fair use flat rate strategies for mobile internet. *Business & Information Systems Engineering, 3*, 269–277. doi:10.1007/s12599-011-0172-6

Gans, J. S., King, S. P., & Wright, J. (2005). Wireless communications. In S. K. Majumdar, I. Vogelsang, & M. E. Cave (Eds.), *Handbook of telecommunications economics* (Vol. 2, pp. 241–285)., Technology Evolution and the Internet Amsterdam: Elsevier.

Gentzoglanis, A., & Henten, A. (2010). *Regulation and the evolution of the global telecommunications industry*. Cheltenham, UK, Northampton, MA: Edward Elgar.

Georg, O. (1996). *Telekommunikationstechnik: Eine praxisbezogene Einführung*. Berlin: Springer.

Gerpott, T. J. (1999). Strukturwandel des deutschen Telekommunikationsmarktes. In D. Fink & A. Wilfert (Eds.), *Handbuch Telekommunikation und Wirtschaft: Volkswirtschaftliche und Betriebswirtschaftliche Perspektiven* (pp. 49–75). München: Verlag Franz Vahlen.

Gerpott, T. J. (2003). Unternehmenskooperationen in der Telekommunikationswirtschaft. In J. Zentes, B. Swoboda, & D. Morschett (Eds.), *Kooperationen, Allianzen und Netzwerke* (pp. 1087–1110). Wiesbaden: Gabler Verlag.

Grishunin, S., & Suloeva, S. (2015). Project controlling in telecommunication industry. In S. Balandin, S. Andreev, & Y. Koucheryavy (Eds.), *Internet of things, smart spaces, and next generation networks and systems* (pp. 573–584). Cham: Springer International Publishing.

Grover, V., & Saeed, K. (2003). The telecommunication industry revisited. *Communications of the ACM, 46*, 119–125. doi:10.1145/792704.792709

Hanna, N. (2010). *Enabling enterprise transformation: Business and grassroots innovation for the knowledge economy, Innovation, technology, and knowledge management*. New York, London: Springer.

IDATE. (2014). *Future telecom: Trends and scenarios for 2025*.

IDATE Research. (2013). *OTT video: Opportunities for telcos around VoD, SVOD and telco CDN*.

Informa Telecoms & Media. (2014). *Industry outlook 2014—Digital futures: Creating new roles and value chains*.

ITU. (1998). *Global information infrastructure principles and framework architecture*. ITU-T recommendation Y.110.

ITU. (2012). *Trends in telecommunication reform 2012—Smart regulation for a broadband world*. Geneva: ITU.

ITU. (2013). *Trends in telecommunication reform 2013—Transnational aspects of regulation in a networked society*. Geneva: ITU.

ITU. (2015a). *Key ICT indicators for developed and developing countries and the world*.

ITU. (2015b). *ICT facts and figures—The world in 2015*.

Jain, R., & Paul, S. (2013). Network virtualization and software defined networking for cloud computing: A survey. *IEEE Communications Magazine, 51*, 24–31. doi:10.1109/MCOM. 2013.6658648

Kendall, P. (2013). *Is VoLTE the answer to the OTT voice threat?*

Kimiloglu, H., Ozturan, M., & Kutlu, B. (2011). Market analysis for mobile virtual network operators (MVNOs): The case of Turkey. *International Journal of Business and Management, 6*. doi:10.5539/ijbm.v6n6p39

Knightson, K., Morita, N., & Towle, T. (2005). NGN architecture: Generic principles, functional architecture, and implementation. *IEEE Communications Magazine, 43*, 49–56. doi:10.1109/ MCOM.2005.1522124

Laudon, K. C., & Traver, C. G. (2015). *E-commerce: Business, technology, society* (11th ed.). Boston: Pearson.

Leung, V. C. M., Lai, R. X., Chen, M., & Wan, J. (2015). In *5th International Conference on Cloud computing, CloudComp 2014*, Guilin, China, October 19–21, 2014 (Revised selected papers).

Lewis, L. (2001). *Managing business and service networks*. New York: Kluwer Academic Publishers.

Li, F., & Whalley, J. (2002). Deconstruction of the telecommunications industry: From value chains to value networks. In *Telecommunications Policy* (pp. 451–472). Amsterdam: Elsevier Science Ltd.

Liu, L., Gu, M., & Ma, Y. (2015). Research on the key technology of M2M gateway. In Q. Zu, B. Hu, N. Gu, & S. Seng (Eds.), *Human centered computing* (pp. 837–843). Cham: Springer International Publishing.

Lyall, F. (2011). *International communications: The international telecommunication union and the universal postal union*. Burlington, VT: Ashgate.

Maitland, C. F., Bauer, J. M., & Westerveld, R. (2002). The European market for mobile data: Evolving value chains and industry structures. *Telecommunications Policy, 26*, 485–504. doi:10.1016/S0308-5961(02)00028-9

Matsuda, F., & Kosaka, M. (2016). Hitachi Construction Machinery Co., Ltd.—M2M and cloud computing based information service. In J. Wang, M. Kosaka, & K. Xing (Eds.), *Manufacturing servitization in the Asia–Pacific* (pp. 75–92). Singapore, Singapore: Springer.

Mell, P., & Grance, T. (2011). *The NIST definition of cloud computing*. Gaithersburg: NIST.

Mikkilineni, R., & Sarathy, V. (2009). *Cloud computing and the lessons from the past* (pp. 57–62). IEEE. doi:10.1109/WETICE.2009.14

Mikkonen, K., Hallikas, J., & Pynnönen, M. (2008). Connecting customer requirements into the multi-play business model. *Journal of Telecommunications Management, 2*, 177–188.

Misra, K. (2004). *OSS for telecom networks: An introduction to network management*. London: Springer.

Mourshed, M., Hediger, V., & Lambert, T. (2014). *Gulf cooperation council health care: Challenges and opportunities*. Riyadh: Gulf Cooperation Council.

OECD. (2014). *Measuring the digital economy*. Paris: OECD Publishing.

OECD. (2015). *Wireless mobile broadband subscriptions*. France: OECD Publishing.

Ovum. (2014a). *Innovative broadband pricing strategies*.

Ovum. (2014b). *Digital operator strategies*.

Park, R. C., Jung, H., Shin, D.-K., Kim, G.-J., & Yoon, K.-H. (2015). M2M-based smart health service for human UI/UX using motion recognition. *Cluster Computing, 18*, 221–232. doi:10.1007/s10586-014-0374-z

Peppard, J., & Rylander, A. (2006). From value chain to value network. *European Management Journal, 24*, 128–141. doi:10.1016/j.emj.2006.03.003

Picot, A. (Ed.). (2006). *The future of telecommunications industries*. Berlin, Heidelberg: Springer-Verlag.

Picot, A. (2007). Tiefgreifende Veränderungen im Ecosystem der Telekommunikationsindustrie. In A. Picot & A. Freyberg (Eds.), *Infrastruktur Und Services—Das Ende Einer Verbindung?* (pp. 13–27). Berlin, Heidelberg: Springer.

Plunkett, J. W. (2014). *Plunkett's telecommunications industry almanac 2015: The only comprehensive guide to the telecommunications industry*.

Pospischil, R. (1993). Reorganization of European telecommunications: The cases of British Telecom, France Télécom and Deutsche Telekom. *Telecommunications Policy, 17*, 603–621.

Pouillot, D. (2013). *Future telecoms: Market scenarios and trends up to 2025*. Idate.

Pousttchi, K., & Hufenbach, Y. (2009). *Analyzing and categorization of the business model of virtual operators* (pp. 87–92). IEEE. doi:10.1109/ICMB.2009.22

Pousttchi, K., & Hufenbach, Y. (2011). Value creation in the mobile market: A reference model for the role(s) of the future mobile network operator. *Business & Information Systems Engineering, 3*, 299–311. doi:10.1007/s12599-011-0175-3

Qian, L., Luo, Z., Du, Y., & Guo, L. (2009). Cloud computing: An overview. In M. G. Jaatun, G. Zhao, & C. Rong (Eds.), *Cloud computing* (pp. 626–631). Berlin, Heidelberg: Springer.

Sale, S. (2013). *OTT communication services worldwide: Stakeholder strategies*.

Sapien, M. (2011). *The enterprise vertical strategies of major telcos*. Ovum.

Snoeck, M., & Michiels, C. (2002). Domain modelling and the co-design of business rules in the telecommunication business area. *Information Systems Frontiers, 4*, 331–342.

STL Partners. (2014). *Five principles for disruptive strategy*.

Tanenbaum, A. S., & Wetherall, D. (2014). In Pearson New internat, Pearson Custom Library (Eds.), *Computer networks* (5th ed.). Harlow, Essex: Pearson Education.

Tardiff, T. J. (2007). Changes in industry structure and technological convergence: Implications for competition policy and regulation in telecommunications. *International Economics and Economic Policy, 4*, 109–133. doi:10.1007/s10368-007-0083-7

Taylor, M. E. (2002). Customer demand analysis. In M. E. Cave, S. K. Majumdar, & I. Vogelsang (Eds.), *Structure, Regulation and Competition: Vol. 1. Handbook of telecommunications economics* (pp. 97–142). Amsterdam: Elsevier.

Telecommunications Industry Association. (2015). *TIA's 2015–2018 ICT market review & forecast.*

TM Forum. (2015). *Information framework (SID): Concepts and principles (GB922),* Version 15.0.0. ed.

Trovati, M., Hill, R., Anjum, A., Zhu, S. Y., & Liu, L. (Eds.). (2015). *Big-data analytics and cloud computing.* Cham: Springer International Publishing.

Velasco-Castillo, E., & de Renesse, R. (2014). *Digital economy readiness index: Mapping telco innovation and digital strategies.* London: Analysys Mason Limited.

Verma, D. C., & Verma, P. (2014). *Techniques for surviving the mobile data explosion.*

Vijayakumar, V., & Neelanarayanan, V. (Eds.). (2016). In *Proceedings of the 3rd International Symposium on Big Data and Cloud Computing Challenges (ISBCC—16'), Smart Innovation, Systems and Technologies.* Cham: Springer International Publishing.

Werner, M. (2010). *Nachrichtentechnik: Eine Einführung für alle Studiengänge.* Wiesbaden: Vieweg+Teubner.

Wieland, R. A. (2007). Konvergenz aus Kundensicht. In A. Picot & A. Freyberg (Eds.), *Infrastruktur Und Services—Das Ende Einer Verbindung?* (pp. 43–67). Berlin, Heidelberg: Springer.

Wu, G., Talwar, S., Johnsson, K., Himayat, N., & Johnson, K. D. (2011). M2M: From mobile to embedded internet. *IEEE Communications Magazine, 49*, 36–43. doi:10.1109/MCOM.2011. 5741144

Wulf, J., & Zarnekow, R. (2011a). Cross-sector competition in telecommunications: An empirical analysis of diversification activities. *Business & Information Systems Engineering, 3*, 289–298. doi:10.1007/s12599-011-0177-1

Wulf, J., & Zarnekow, R. (2011b) How do ICT firms react to convergence? An analysis of diversification strategies. In *ECIS 2011 Proceedings.* Paper 97.

Yahia, I. G. B., Bertin, E., & Crespi, N. (2006). Next/new generation networks services and management. In *Proceedings of the International Conference on Networking and Services, ICNS'06* (p. 15). Washington, DC, USA: IEEE Computer Society. doi:10.1109/ICNS.2006.77

Zeithaml, V. A., & Bitner, M. J. (2003). *Services marketing: Integrating customer focus across the firm.* Boston: McGraw-Hill Education.

Zheng, K., Fanglong, H., Wang, W., Xiang, W., & Dohler, M. (2012). Radio resource allocation in LTE-advanced cellular networks with M2M communications. *IEEE Communications Magazine, 50*, 184–192. doi:10.1109/MCOM.2012.6231296

Chapter 3
Understanding the Methodical Principles

Abstract Understanding the methodical principles is indispensable for the successful adaptation of organization, processes, data, and applications to the changed industry conditions. In most cases, those adjustments are related to the various different parts of a telecommunications operator. The planning, design, and realization of those changes are a complex endeavor which, in most situations, takes several years, involves huge project teams, and impacts major parts of the enterprise. Without clear structures and guidelines, the risk of inconsistent and singular solutions is high. The overriding challenge is to understand the interrelations between the different enterprise parts and take decisions that are beneficial from the overall enterprise perspective. The general methodical foundation of the solution design is related to information systems modeling. In this context, information systems are a complex construct comprised of employees, their organizational responsibilities, their activities that create the enterprise's outcome, as well as applications that support and automate activities. Enterprise architectures provide a general structure to plan, design, and implement those complex solutions. Content-wise, reference models are used as recommendations. In the telecommunications industry, the TM Forum offers well-accepted reference models for processes, data, and applications. From the dynamic perspective, concepts of enterprise architecture management and enterprise transformation support the planning and implementation. In this chapter, a general introduction to information systems modeling (cf. Sect. 3.1), a description of enterprise architecture approaches (cf. Sect. 3.2), reference modeling (cf. Sect. 3.3), relevant reference models for the telecommunications industry (cf. Sect. 3.4), and enterprise transformation (cf. Sect. 3.5) are presented.

The objective of this book is to support the transformational needs of telecommunications operators. From a methodical viewpoint, this requires designing a solution that solves the specific requirements of telecommunications operators, and

© Springer International Publishing AG 2017
C. Czarnecki and C. Dietze, *Reference Architecture
for the Telecommunications Industry*, Progress in IS,
DOI 10.1007/978-3-319-46757-3_3

implementing this solution. The solution itself should cover all relevant areas of a telecommunications operator. As a first step, it is necessary to identify all relevant parts and their interrelations, which can be understood as the *structural perspective*. In the second step, concrete and specific solutions should be developed, which can be seen as the *topical perspective*. Finally, telecommunications operators would only benefit from the solution design after all changes have been applied to their current situation, which is the *implementation perspective*.

From a methodical viewpoint, information system research offers the following concepts to support the solution design and implementation:

- *Enterprise Architecture Management* (EAM) provides concepts to describe and change the fundamental parts of an enterprise according to different layers (e.g., strategy, processes, data) (Winter and Fischer 2007, p. 7).
- *Reference Modeling* covers the design and usage of concrete solution blueprints that are provided as a recommendation for a certain problem domain (Fettke and Loos 2007a, pp. 3–5).
- *Enterprise Transformation* describes the fundamental change of an enterprise in order to react, for example, to market or technological changes (Rouse and Baba 2006).

Those three concepts are used to support the structural, content, and implementation perspectives (cf. Fig. 3.1). Through Enterprise Architecture Frameworks, EAM provides a structure for the identification and description of the relevant parts of a telecommunications operator. The concrete representation is an enterprise architecture. While the structure is provided by the Enterprise Architecture Framework, the design of the specific solution requires recommendations with regards to the content. This content is given by reference models. The *TM Forum Frameworx* offers industry-specific reference models for processes, data, and applications. In contrast, ITIL is a reference model for the functional domain *IT service management* and not dependent on a specific industry. The implementation of the developed solution impacts the concrete structures and systems of an enterprise based on the conceptual solution design. In most cases, the roll-out of software products is part of the implementation. In the telecommunications industry, most vendors assure conformance between their software products and TM Forum Frameworx. This book stays on the conceptual level and does not discuss specifics of concrete software products. The overall change of an enterprise from its current to a targeted future state is supported by enterprise transformation. The conceptual solution design is related to information systems modeling (cf. Sect. 3.1). In the following, a detailed description of enterprise architecture approaches (cf. Sect. 3.2), reference modeling (cf. Sect. 3.3), relevant reference models for the telecommunications industry (cf. Sect. 3.4), and enterprise transformation (cf. Sect. 3.5) is presented.

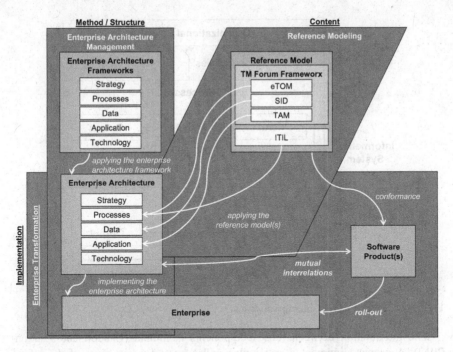

Fig. 3.1 Overview of methodical concepts

3.1 Fundamentals of Information Systems Modeling

In general, the context of this book is related to information systems modeling. Therefore, a short introduction to this topic is provided. The modeling of information systems is an important aspect of their development (Avison and Fitzgerald 2006, pp. 109–118; Satzinger 2015, pp. 58–59; Stair and Reynolds 2012, pp. 26–30).

An *information system* supports business tasks through the processing of information (Stair and Reynolds 2012, p. 8). It consists of various components that are interrelated with each other (Satzinger 2015, p. 4). Besides technical components, an information system includes the people operating the technical components or performing manual tasks (Satzinger 2015, p. 4). An information system can be structured into technology, organization, and management (Laudon and Laudon 2012). Therefore, an information system is more than a technical solution provided by a software program (Laudon and Laudon 2012). Both processes and applications transform an input into an output (Stair and Reynolds 2012, p. 8). The application supports or automates processes by a partial or complete conducting of certain activities through technical component(s). It also enables new or innovative processes that were not possible without technical support. Applications are realized through software and hardware. Processes are related to the organizational structure defining responsibilities. Overall, an information system requires the consideration of organizational structure, processes, and applications (cf. Fig. 3.2).

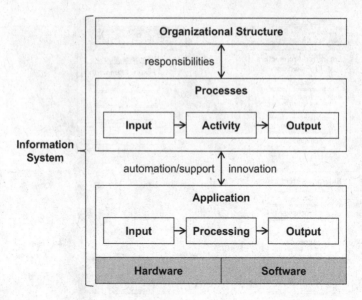

Fig. 3.2 Components of an information system[1]

The interrelation between those components is mutual (Laudon and Laudon 2012). A proper alignment is a complex endeavor and a key topic of the information systems discipline. In this context, the documentation or design of information systems is an important step (Avison and Fitzgerald 2006, pp. 109–118; Satzinger 2015, pp. 58–59; Stair and Reynolds 2012, pp. 26–30) that is supported by models. A *model* can be understood as "a representation or abstraction of some aspect of a system" (Satzinger 2015, p. 58). Please see, e.g., Stachowiak (1973), Ludewig (2003), Schalles (2012), Becker und Schütte (2004), or Thomas (2005) for further discussions of the terms *model* and *modeling*. The model development is typically iterative, using textual or graphical modeling languages (Satzinger 2015, pp. 58–59). A modeling language is defined by syntax, semantics, and representation (Harel and Rumpe 2004, pp. 65–68; Karagiannis and Woitsch 2010, p. 631) that can be formalized in a meta modeling language (e.g. Engels et al. 2000; Karagiannis and Woitsch 2010, p. 631; Kühne 2006; Kurpjuweit and Winter 2007). In the context of information systems, the purpose of a model is to describe how the information system works or should work (Rummler and Ramias 2010, p. 84).

The person(s) constructing a model (i.e., modeler) are in most cases different from the person(s) using the model (i.e., model user) (Becker and Schütte 2004). The model users are normally not a uniform group, but rather consist of different target groups, e.g., the business owner and the system implementer. According to

---

[1]The figure is an own illustration. The different components and their interrelation are discussed and illustrated by various publications (Alpar et al. 2014, p. 24; Becker and Schütte 2004, p. 33; Ferstl and Sinz 2008, p. 4; Laudon and Laudon 2012; Satzinger 2015, p. 4; Stair and Reynolds 2012, p. 8) which were used as basis here.

the different modeling purposes and the expectations of the intended target audience, a concrete model provides a sample compared with the real object. In addition, it can be assumed that the modelers have their own background and perceptions which might influence the modeling result. In order to understand the modeling challenges in reality, it is important that a model is not an unbiased illustration of an existing or future object, but the result of a construction influenced by the involved persons (vom Brocke 2003, p. 11). .

The value of a model is, amongst other things, related to its correctness, relevance, and clarity (Rosemann 2003, pp. 59–60). There is a formal correctness of a model that is related to the modeling language, modeling method, and additional modeling conventions (e.g., company-specific rules) (Rosemann 2003, p. 59). Please see Harel and Rumpe (2004) for a differentiation between syntax, semantics, and representation in this context. Based on the modeling language and tools, there is, to some extent, a possibility to prove this formal correctness. At the end the modeling result should support the understanding, development, or implementation of an information system. Besides the formal correctness, the correct model content describing all relevant parts of the existing or future object is crucial; this can be called "fitness for use" (Rosemann 2003, p. 42). The correctness of the model content cannot be proven. It is a construction result as a joint effort of modeler(s) and model users.

In practice, the correct understanding and modeling of the relevant parts of an information system are an important prerequisite for its successful implementation. Therefore, the impact of the explained modeling fundamentals is summarized in Fig. 3.3. The purpose of any modeling is the creation of a concrete model that is somehow related to a concrete enterprise. This interrelation could have different purposes, such as the documentation of the as-is situation, the optimization of existing structures, or the definition of a target state for a planned implementation (Rosemann 2003, pp. 42–47). The concrete model contains a model content that should support the specific modeling purpose—for example, improved procurement activities to realize a process optimization. The modeling itself follows a modeling method in order to formalize the model content in a modeling language. Both modeling method and language could influence the model content.

Typically, various personnel are involved in this modeling. Those personnel have different roles and responsibilities, such as managing specific processes, being responsible for certain IT systems, or being accountable for certain results. According to their concrete roles and responsibilities their involvement in the modeling could vary from active design of certain model parts to formal acceptance of the final results. Each of the involved persons has an own notion of the model, which might be influenced by his own ideas, background, and roles and responsibilities. This notion of the model might differ from the notion of other individuals, and could impact the concrete model. This part is one of the most critical aspects of modeling in a practical context. The finished concrete model is a consensus of the involved persons. The way to achieve this consensus should be carefully planned and defined from the beginning.

Furthermore, the concrete model can be designed by applying (a) reference model(s). In such a case, the reference model is also a result of a modeling activity. In most cases, the reference model is a consensus of various involved persons—i.e.,

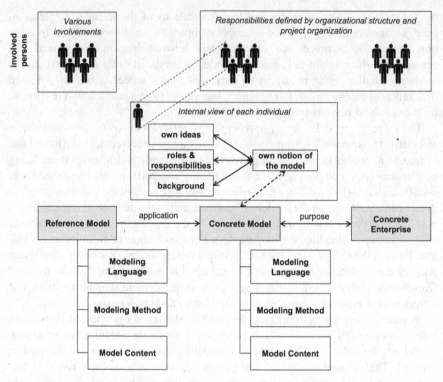

Fig. 3.3 Summary of modeling fundamentals[2]

a standardization committee. Based on the acceptance of the reference model in the intended usage domain, the consensus during the design of the concrete model could be supported by the reference model. On the other hand, the application of a reference model is again a modeling activity.

3.2 Enterprise Architecture

The modeling of information systems is supported by concepts of enterprise architectures. In practice, the development, optimization, or implementation of information systems is a complex task that influences various parts of an enterprise. The modeling of those parts is supported by enterprise architecture concepts. They provide structures, templates, and methods for the overall design and description of an enterprise. Typically, those concepts are generic and independent from concrete

---

[2]The figure is an own illustration. The theoretical fundamentals of modeling and models are based on the various publications discussed in this section. Especially the construction-based modeling illustrated by Schütte (1998, p. 61) and the discussion of the term *modeling* by Thomas (2005) were used as input.

industries. From a methodical perspective, enterprise architectures can be seen as a structural element that supports the overall solution design.

Section 3.2.1 starts with a definition of relevant terms and an introduction of the general concept of enterprise architectures. An important differentiation is a concrete enterprise architecture as a representation of a specific situation. It is supported by Enterprise Architecture Frameworks, which can be seen as an empty structure that is filled with content to create a concrete enterprise architecture. Enterprise Architecture Frameworks are explained in Sect. 3.2.2. The dynamic aspect of designing, implementing, maintaining, and optimizing an enterprise architecture is part of Enterprise Architecture Management and discussed in Sect. 3.2.3. The Open Group Architecture Framework (TOGAF) is an example of an Enterprise Architecture Framework that is briefly introduced in Sect. 3.2.4.

3.2.1 Introduction to Enterprise Architecture

Enterprise architectures are widely used for the modeling of information systems (e.g. Ahlemann 2012; Schekkerman 2004; Van Den Berg and Van Steenbergen 2006; Winter and Sinz 2007). Zachman can be seen as a pioneer of enterprise architectures. He defined them as a "set of design artifacts, or descriptive representations, that are relevant for describing an object such that it can be produced to requirements (quality) as well as maintained over the period of its useful life (change)" (Zachman 1997). In practice, the scope of enterprise architectures is normally broad, and often covers the whole or major parts of the enterprise (Winter and Fischer 2007). Therefore, there is not one single representation of a concrete architecture, but various different views according to purpose and target audience (Zachman 1997). This concept is comparable with the construction plans in traditional engineering disciplines (Aier et al. 2008, p. 118), which might include different plans for statics and for electrical cabling.

According to the ANSI/IEEE Standard 1471-2000, Enterprise Architecture is defined as follows (Winter and Fischer 2007, p. 7): "The fundamental organization of an enterprise embodied in its components, their relationships to each other, and to the environment, and the principles guiding its design and evolution" (according to IEEE Computer Society 2000, p. 3).

For this reason, an Enterprise Architecture includes both the content of the described architecture and procedures for their design and management. From the content perspective, an enterprise architecture is a fundamental and complete illustration of an enterprise (Schekkerman 2004, p. 13). Its purpose is the alignment between strategic and operational as well as business and technical aspects (Aier et al. 2011, p. 645; Schekkerman 2004, p. 13). Accordingly it combines various different artifacts, such as organizational structure, data structure, or network infrastructure. The challenge of an enterprise architecture is the coordination of all relevant aspects and interrelations in a holistic way, so as to allow an understanding of the essential elements and their functioning (Schekkerman 2004, pp. 13–15).

Therefore, enterprise architectures are typically structured into different views, perspectives, or layers that could consist of different models (IEEE Computer Society 2000, p. 5).

For the usage of enterprise architectures in a practical context, it is important to distinguish between the following terms:

- An *Enterprise Architecture* is the construction result for a specific enterprise. It contains concrete models and procedures. Their purpose is a fundamental illustration of the enterprise objects with a broad scope and definition of interrelations (Aier et al. 2009, p. 39; Winter and Fischer 2007, p. 7).
- An *As-Is Enterprise Architecture* describes the current state of the enterprise (Schekkerman 2004, p. 22). It can be used as a starting point or baseline.
- A *To-Be Enterprise Architecture* describes the target state of the enterprise that should be realized in the future through enterprise transformation (Schekkerman 2004, p. 23).
- An *Enterprise Architecture Framework* provides a generic guidance for the development of a concrete Enterprise Architecture. It contains templates, patterns, methods and definitions (Winter and Fischer 2007, p. 7). Thus, it is a structural and methodical blueprint for an Enterprise Architecture.
- *Enterprise Architecture Management* (EAM) focuses on the methodical part of the design, operations, and maintenance of an Enterprise Architecture (Aier et al. 2011, p. 645). Therefore, it includes generic procedures and guidelines. In most cases, it is part of an Enterprise Architecture Framework. Nonetheless, independent EAM approaches are also available (Aier et al. 2011, p. 646).

Finding an appropriate level of detail is a major challenge of enterprise architectures (Zachman 1997). Certainly the complexity of an enterprise does not allow their complete illustration. A holistic view covering the interrelations between all relevant elements is the major purpose of enterprise architectures (Schekkerman 2004, pp. 13–15). The enterprise architecture requires a broad view of the enterprise (Aier et al. 2009, p. 39; Winter and Fischer 2007, p. 7) which abstracts key information from the complex whole in order to focus on the essential elements and their interrelations. Those interrelations are especially important. They support the holistic evaluation of decisions—for example, the impact of changed sales processes on underlying systems and data. According to Aier et al. (2009, p. 39), elements that include only those details relevant for the implementation of a single artifact should most likely not be part of the enterprise architecture. There should be a differentiation between artifacts that influence various elements of the enterprise, and those that are independent (Aier et al. 2009, p. 40). Therefore, we distinguish between the enterprise architecture and further detailed solutions (Aier et al. 2009, p. 40; Schekkerman 2004, pp. 23–24). Those detailed solutions could be again described in architectures—for example, a software architecture linking detailed requirements with software structure and products (Schekkerman 2004, p. 23). The enterprise architecture can be seen as a layered view on essential elements of an enterprise linked to detailed solutions (cf. Fig. 3.4).

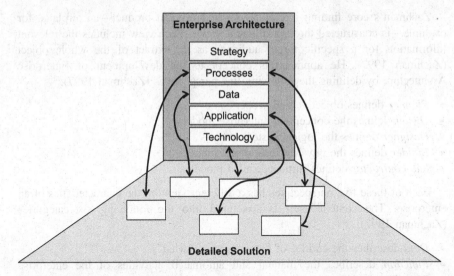

Fig. 3.4 Interrelation between Enterprise Architecture and detailed solutions (according to Aier et al. 2008, p. 39; Schekkerman 2004, p. 24)

In the following sections, there is a description and comparison of different approaches for Enterprise Architecture Frameworks (cf. Sect. 3.2.2) and EAM (cf. Sect. 3.2.3). Both support the definition, design, and management of concrete enterprise architectures by providing generic structures and methods.

3.2.2   Enterprise Architecture Frameworks

Enterprise Architecture Frameworks (EAF) are defined as generic structures and methods that can be used for the construction of a concrete Enterprise Architecture (Schekkerman 2004, p. 85). Their usage in the context of information systems development is well recognized (e.g. Buckl et al. 2009; Saat et al. 2010; Urbaczewski and Mrdalj 2007; Winter and Fischer 2007).

The *Zachman Framework for Enterprise Architecture* is the first serious EAF (Noran 2003, p. 105; Ostadzadeh et al. 2007, p. 375; Schekkerman 2004, p. 89; van't Wout et al. 2010, p. 162). It was published by Zachman (1987), who is recognized as a pioneer of Enterprise Architectures. His work is an interdisciplinary combination of concepts from traditional disciplines of architecture, engineering, and manufacturing. His ideas are helpful for a general understanding of Enterprise Architectures, and are therefore shortly summarized in the following paragraphs. However, there is only a minor practical relevance of the Zachman Framework for the telecommunications industry.

Zachman's core finding is that a complex physical product—an airplane, for example—is constructed through different views. Each view includes the relevant information for a specific target audience as an extract of the whole object (Zachman 1997). He applies this concept to the development of Enterprise Architecture by defining the following five perspectives (Zachman 1997):

- *Planner* defines objective and scope.
- *Owner* defines the conceptual enterprise model.
- *Designer* defines the logical system model.
- *Builder* defines the physical technology model.
- *Sub-contractor* defines further detailed models.

Each of these five perspectives has a different view on the characteristics of an enterprise. The content itself is structured into the following six categories (Zachman 1997):

- *Data* describes the entities of the enterprise ("what").
- *Function* describes the manual and automated activities of the enterprise ("how").
- *Network* describes the physical representation of the enterprise ("where").
- *People* describes the human representation of the enterprise ("who").
- *Time* describes the scheduling of the enterprise ("when").
- *Motivation* describes the objectives and rules of the enterprise ("why").

The general assumption of Zachman (1987) is that each of the five perspectives has a different but relevant content for each of the six categories. For this reason, his framework consists of a matrix of five perspectives and six content categories, which together result in 30 elements. For example, the perspective planner contains a list of processes in the category function, while the designer includes an application architecture in the same category. However, the framework does not include a detailed guidance for the development of these 30 elements. In fact, Zachman (1997) points out that he provides a generic structure which can be used for the development and description of any concrete object.

In the 1990s, various governmental organizations and private companies started the development of EAFs, such as the *Federal Enterprise Architecture Framework* (FEAF) or the *Integrated Architecture Framework* (IAF) (Schekkerman 2004, p. 105). In 1995 the non-profit organization *The Open Group* developed *The Open Group Architecture Framework* (TOGAF). Please see Table 3.1 for an overview of selected EAF.

Table 3.1 Overview of selected EAF (according to Schekkerman 2004, pp. 89–141; Urbaczewski and Mrdalj 2007, pp. 18–19)

Name	Background	Content
Zachman framework for enterprise architecture	Published by Zachman (1987)	Five perspectives: (1) planner, (2) owner, (3) designer, (4) builder, (5) sub-contractor Six categories: (1) data, (2) function, (3) network, (4) people, (5) time, (6) motivation
Federal enterprise architecture framework (FEAF)	Version 1.1 was published in 1998 by the USA Chief information officer council, based on it the development of the federal enterprise architecture (FEA) was started in 2002	Four architecture levels: (1) business, (2) data, (3) application, (4) technology The methodology consists of architecture drivers, strategic direction, current architecture, target architecture, transitional processes and standards
Treasury enterprise architecture framework (TEAF)	Published in 2000 by the US department of the treasury and based on FEAF	Four architecture levels: (1) function, (2) information, (3) organization, (4) infrastructure. The methodology contains the definition of an architecture strategy, an architecture management process and a repository
Integrated architecture framework (IAF)	Developed in 1996 as a proprietary framework by Capgemini, further influenced by the merger with Ernst and young consulting in 2001	Definition of the following mandatory content for an integrated architecture: (1) business/organization, (2) information (incl. information flows and relations), (3) information systems, (4) technology/infrastructure
The open group architecture framework (TOGAF)	Developed in 1995 by *The Open Group* based on the *technical architecture framework for information management* (TAFIM) from the US department of defense, today version 9.1 exists	The *Architecture Development Method* (ADM) contains the methodology of a concrete architecture. The *Architecture Content Framework* provides templates and structures for the architecture content
Extended enterprise architecture framework (E2AF)	Developed in 2002 by the *institute for enterprise architecture developments* (IFEAD) based on Zachman framework, IAF and FEAF	Six levels of abstraction: (1) contextual, (2) environmental, (3) conceptual, (4) logical, (5) physical, (6) transformational Four aspects: (1) business, (2) information, (3) information systems, (4) technology/infrastructure

The table is a translated version based on Czarnecki (2013, p. 20)

There is obviously not one general preference of an EAF, but various different EAFs. Due to the historical interrelations of their development, some similarities of their structure can be observed (Schekkerman 2004, p. 89). On the other hand, the available EAFs are to some extent quite different in their logic and scope (Urbaczewski and Mrdalj 2007, p. 22). In fact, a general and widely accepted EAF is not available (Winter and Fischer 2007, p. 7). Therefore, a real-life project for the development of a concrete enterprise architecture would typically start with the selection of an appropriate EAF. For a general understanding of the scope of a specific EAF, the study of Winter and Fischer (2007, p. 8) is helpful. They have analyzed various EAFs, and found out that the following perspectives are covered in most EAFs (Winter and Fischer 2007, p. 8):

- The *business perspective* describes the fundamental aspects of the organization from a strategic viewpoint, such as market segments and product offerings.
- The *process perspective* contains such aspects as business processes, responsibilities, performance indicators, and information flows.
- The *integration perspective* structures the components of information systems—for example, through application categories or data flows.
- The *software perspective* describes artifacts relevant for the definition of software, like software services and data structures.
- The *technology and infrastructure perspective* covers the required hardware for the relevant information and communication systems.

The objective of an enterprise architecture is a fundamental description of an enterprise (Winter and Fischer 2007, p. 7). This does not mean that it always covers all parts of an enterprise. It is also usual to limit the scope based on a specific project to those parts that are relevant for the implementation of a new software system. Defining the relevant scope is an important prerequisite for the selection and application of an EAF. Zachman (1997) mentioned that the appropriate granularity might be a challenge during the development of an enterprise architecture. He proposed a hierarchical structure of architectures which is comparable with the bill of material used to define complex physical products. In contrast to a specific solution model (e.g., a data model for new customer database), the purpose of an enterprise architecture is to provide different perspectives and to illustrate interrelations between different objects. Winter und Fischer (2007, p. 12) argue that an enterprise architecture should be broad rather than deep. They propose that the different perspectives should follow a hierarchical structure. On a higher level, the big picture of the architecture is illustrated and further detailed to a certain extent. For more details, they propose interfaces to specialized architectures covering only a limited scope (Winter and Fischer 2007, p. 13).

In fact, projects sometimes suffer from the late or missing involvement of the relevant topics. As an example, it is possible that a CRM implementation project is mainly driven by the IT Department, which then might focus mainly on the system functionalities and their technical advantages. In this scenario, the involvement of the business side, their requirements, and a linkage to the overall strategic targets are most likely not considered. The advantages of an EAF already start in the supporting the definition and scoping of a concrete project. In addition, an EAF also provides methodical guidance, which is part of the Enterprise Architecture Management (EAM) discussed in the next section.

3.2.3 Enterprise Architecture Management

A concrete enterprise architecture describes the relevant artifacts of an enterprise from an overall viewpoint—i.e., it consists of different models defining specific content (e.g., a process model detailing the activities within an enterprise). The development of such architectures is supported by Enterprise Architecture Frameworks (EAF) that includes templates, structures and methods. Those methods are then part of the Enterprise Architecture Management (EAM). The scope of EAM goes beyond the pure modeling. It covers the planning and controlling of enterprise-wide architectural changes (Aier et al. 2011, p. 645).

The coverage of EAM tasks by a specific EAF is highly dependent on the concrete EAF. For example, TOGAF includes the *Architecture Development Method* (ADM) that provides a broad method for EAM. In contrast, the Zachman Framework (Zachman 1987) is purely focused on the structure of the architectural content. A real-life project requires both the architectural structure and guidance for the EAM tasks. Depending on the selection of a specific EAF, the EAM approach varies significantly (Schekkerman 2004, p. e.g.; Urbaczewski and Mrdalj 2007; van't Wout et al. 2010). Therefore, this section gives a short introduction to EAM from a general perspective.

Aier et al. (2011, p. 646) summarize the typical EAM tasks as follows:

1. Strategic design of an architecture vision
2. Development and maintenance of the as-is architecture models
3. Development and maintenance of the to-be architecture models
4. Migration planning
5. Architecture implementation
6. Analysis of the architecture based on architecture models
7. Communication of architecture guidelines and principles

This summary of EAM tasks shows clearly the importance of dynamic aspects. A purely static definition of an enterprise architecture would be without major benefits. Typically, enterprise architectures are used in transformation or implementation projects. Hence, the change of the current situation towards a target picture is a primary objective. An indispensable prerequisite, therefore, is a clear distinction between the as-is and the target architecture. In fact, this is a frequently observed shortfall in real-life projects. To some extent, there is a high risk that a modeler might include potential improvements in the as-is architecture or considers the limitations of the current situation in the target picture. Without this clear distinction, the migration planning and architecture implementation are a challenging endeavor.

Another critical success factor is the involvement of all relevant stakeholders. A target architecture combines business and technical artifacts (e.g. Ahlemann 2012, p. 61; Aier and Gleichauf 2010b), and should therefore combine stakeholders from business and technical departments. Consequently, the organizational positioning of EAM is an important success factor. From a historical perspective, the roots of EAM can be found in initiatives related to computer science (Schekkerman 2004, pp. 89–141; Urbaczewski and Mrdalj 2007, pp. 18–19). The purpose of EAM is a holistic management of essential artifacts of an enterprise, which is certainly a strategic undertaking. Ahlemann (2012, p. 101) points out that, in practice, most EAM initiatives are related to the CIO organization. The minority, directly reporting to the CEO, could achieve higher alignment and acceptance. Also Saat et al. (2010, p. 15) describe the advantage of EAM for the decision support of technical and business-oriented target groups of an enterprise. This viewpoint is supported by an empirical study of Aier et al. (2011). They analyzed the following three different usage scenarios for EAM: (1) only business focus, (2) only IT focus, (3) balanced focus between business and IT. The result of their study shows that a higher benefit is realized by an EAM usage with a balanced focus between business and IT (Aier et al. 2011, p. 653). For that reason, understanding EAM as an interdisciplinary and cross-functional task, which is definitely not the sole responsibility of the IT department alone, is recommended. In fact, an early involvement of both technical and business departments is a critical success factor for any EAM initiative.

3.2.4 The Open Group Architecture Framework

As described in the previous sections, several different structures and methods for enterprise architectures are available. The selection of a concrete Enterprise Architecture Framework (EAF) is dependent on various criteria and might be influenced by strategic decisions and regulatory obligations. Hence, this book does not provide a recommendation for a single EAF. However, for the understanding of the general concept of enterprise architecture, a more detailed explanation of a concrete EAF is helpful. As a concrete EAF, *The Open Group Architecture Framework* (TOGAF) is used as methodical guidance in the context of the TM Forum reference models (cf. Sect. 3.4). Therefore, in this section a short summary of TOGAF is provided.

TOGAF was developed in 1995 by *The Open Group* based on the *Technical Architecture Framework for Information Management* (TAFIM) from the US Department of Defense. It is independent of a concrete industry, and developed by a neutral consortium. Today TOGAF exists in version 9.1 published by The Open Group (2011). In addition, numerous publications provide a general overview of TOGAF and discuss its application (e.g. Desfray and Raymond 2014; Greefhorst and Proper 2011, pp. 181–185; Perks and Beveridge 2003, pp. 76–94).

The general structure of TOGAF differentiates between the *Architecture Development Method* (ADM) containing the methodology of a concrete architecture and the *Architecture Content Framework* providing templates and structures for the architecture content.

TOGAF ADM is structured into the following phases (The Open Group 2011):

- *Preliminary* covers the definition of the organizational scope and the evaluation of the architectural maturity of the enterprise.
- *Phase A: Architecture Vision* defines a vision (business values and capabilities) to be delivered by the planned enterprise architecture. In this phase, the architecture project is established including scope statement, principles, value proposition, and responsibilities.
- *Phase B: Business Architecture* covers the development of a target business architecture (product strategy, processes, organization) aligned with the defined business objectives. This phase starts with a baseline (as-is architecture), develops a target architecture, and performs a gap analysis. It also includes the selection of relevant reference models for the concrete business modeling.
- *Phase C: Information Systems Architecture*[3] includes the development of a target architecture for data and applications in order to realize the business architecture and business objectives. Both the data and application architecture follow similar steps as the business architecture: selection of relevant reference models, baseline (as-is architecture), target architecture, and gap analysis.
- *Phase D: Technology Architecture* defines the technical target architecture to realize the logical as well as physical data and application components.
- *Phase E: Opportunities and Solutions* covers the complete planning of the architecture implementation. For an incremental implementation, the required transition architectures are also identified.
- *Phase F: Migration Planning* includes the finalization of the overall planning as well as the coordination of the implementation and migration plan.
- *Phase G: Implementation Governance* covers the steering of implementation projects with a clear focus on the conformance to the target architecture.
- *Phase H: Architecture Change Management* includes the overall lifecycle of all required architectural changes.
- *Requirements Management* assures a continuous and structured management of requirements with respect to all ADM phases.

For each phase, TOGAF provides a detailed description of input, activities, and output. With the input and output, the methodology is linked to the content structure, which is further detailed by the *Architecture Content Framework*. Its objective is structural guidance for a consistent definition and presentation of architectural content, which is structured into the following three categories (The Open Group 2011):

- *Deliverables* are the formally specified and reviewed output of projects—e.g., a concrete target business architecture as output of phase B.

---

[3]Please note that the term *information systems* used in TOGAF differs from the usage in this book which understands information systems as a superordinate that includes, amongst others, processes and applications (cf. Sect. 3.1).

- *Artifacts* are part of a deliverable and describe a specific aspect of an architecture. They are either catalogs (listing different items), matrices (illustrating
  relations), or diagrams (graphical illustration of items).
- *Building blocks* are architectural capabilities that can be combined with other
  building blocks. They are described by artifacts. A building block could consist
  of many artifacts and artifacts could belong to many building blocks.

For all building blocks, TOGAF provides a content meta model that is structured
into (1) architecture principles, vision and requirements, (2) business architecture,
(3) information systems architecture, (4) technology architecture, and (5) architecture realization (The Open Group 2011). In addition, further guidance for the
definition of artifacts and deliverables is provided. Furthermore, TOGAF provides
guidelines and tools for the management of different enterprise architectures
(Enterprise Continuum and Tools), generic concepts for reference models as well as
guidelines for the set-up and operations of a architecture function within an
enterprise (Architecture Capability Framework) (The Open Group 2011).

TOGAF can be seen as an extensive EAF that provides detailed guidance for
structuring the architectural content as well as managing the architecture development and realization. TOGAF follows a clear hierarchical logic that is comparable with the five perspectives proposed by Winter and Fischer (2007, p. 8):
business, processes, integration, software, and technology (cf. Sect. 3.2.2). Due to
the comprehensive scope of TOGAF, it is complex and requires customization
when it is used in a real-life project. In addition, further guidance regarding the
content itself is required. Therefore, reference models are typically used (cf.
Sect. 3.3) in combination with TOGAF.

3.3 Reference Modeling

From a content perspective, *reference models* offer recommendations for the
modeling of information systems. They give a blueprint for the design of a concrete
model that is customized according to specific requirements. The challenge of
reference models is that they should fit to a wide range of usage scenarios but also
be specific enough for a concrete problem domain. Typically, reference models are
specific for certain industries or functions. Their content could cover all different
parts of an information system including processes, data, and applications. In
practice, the usage of reference models is related to cost savings and quality
improvements. Section 3.3.1 provides a definition of relevant terms and an introduction of reference modeling. A categorization of the broad range of different
reference models is discussed in Sect. 3.3.2. The development and usage of reference models are explained in Sect. 3.3.3.

3.3.1 Introduction to Reference Modeling

Reference models are widely accepted tools to support the development of information systems (Becker et al. 2003; Becker and Delfmann 2007; Fettke and Loos 2007b; Thomas 2006a). Various accepted reference models exist for different industries or functions (e.g. Bolstorff and Rosenbaum 2012; Orand 2013; Scheer 1998). In practice, the development of enterprise architectures (cf. Sect. 3.2) is supported by reference models, for example as proposed by TOGAF (Lankhorst 2013, p. 24).

A reference model can be defined as follows[4]: "a reference model [...] is an information model used for supporting the construction of other models "(Fettke and Loos 2007a, pp. 3–4; Thomas 2006a, p. 491). The advantages of using a reference model are higher efficiency and quality. A concrete development (a so-called application model) is based on existing results that are used as a reference or blueprint (vom Brocke 2007, p. 49). The concept of re-using reference models is comparable with design patterns in computer science (Winter et al. 2009).

Reference modeling describes the construction and usage of reference models (Fettke and Loos 2007a, p. 5; vom Brocke 2007, pp. 48–51). The reference model itself provides the topical results, while the development and application method is covered in the reference modeling. Although the usage of reference models is widely accepted, a consensus about general methods for their development, evaluation, and usage have yet to emerge (Fettke and Loos 2007a, pp. 9–11; Frank 2007, p. 122). The major challenge is that the reference model should, on the one hand, be general enough to be applicable for different situations and, on the other hand, concrete enough to be helpful in an implementation. This is summarized in the requirement that a reference model should be generalized and have a recommendatory character (Fettke and Loos 2004a). In addition, the development of a reference model is typically separated from its usage. The reference model developer must cope with the challenge of developing a useful recommendation for an unknown model user (vom Brocke 2007, p. 49).

In a practical context, the acceptance of a reference model is typically based on a consensus governed by a neutral authority. The decision about the usage of a reference model is then comparable with a sales decision in a business environment which is influenced by hard factors (e.g., price and topical fit) as well as soft factors (e.g., trust in the reference model developer) (Frank 2007). Furthermore the usage of a reference model is dependent on its diffusion in that specific problem domain. A widely used reference model could become a de facto standard—for example, the IT Infrastructure Library (ITIL) for IT service management (e.g. Orand 2013).

---

[4]There are various discussions about the definition of the term reference model in the scientific community. Please see e.g. Becker and Delfmann (2007) and Fettke and Loos (2007b) for further details.

3.3.2   Types of Reference Models

The general idea of providing a reference for the development of a concrete model leads to a broad variety of usage scenarios for reference models. As early as 1979, the *Reference Model of Open Systems Interconnection* provided a reference structure for the communication in an open system environment (Zimmermann 1980, p. 425), which is still widely accepted today. There are various other examples for specific industries, functions, or IT systems (e.g. Bolstorff and Rosenbaum 2012; Orand 2013; Scheer 1998). The concept of reference models is not limited to information systems. A reference model exists, for example, for the development of a business model (Osterwalder and Pigneur 2010) or a reference model for the role of mobile operators (Pousttchi and Hufenbach 2011). These examples show that concrete reference models could extensively vary in scope and content.

From a scientific perspective, a first criterion for differentiating reference models is the extent to which they are described and observed (Fettke and Loos 2004a, pp. 332–333). This book focuses on reference models which are relevant in a practical context. It is assumed that these reference models are clearly described and labeled for being re-used. In a practical context, detecting the existence of a reference model is also not always clear. The terminology is often mixed with terms such as *blueprint*, *best practice* and *framework*. The relevant reference models for the telecommunications industry are discussed in Sect. 3.4.

A reference model is a general model that is re-usable in certain usage scenarios (Fettke and Loos 2007a, pp. 3–4; Thomas 2006a, p. 491). The assumption is that for similar problems, a similar solution could be applied. A similarity of these usage scenarios is an important prerequisite for a successful re-use. Therefore, the usage domain of the reference model is an important criterion to differentiate reference models. The usage domain could be either a certain industry (retail, insurance, telecommunications) or functional focus (logistics, IT service management, communication of open systems) (Fettke and Loos 2004b, p. 23).

Defining re-usable content for a certain usage domain is a promising concept in various different areas. The scope and content of reference models vary from broad reference information models (Scheer et al. 2007) to more restricted reference process models (Fettke et al. 2006, p. 1; Krcmar 2005, p. 125). The topical scope of a reference model is another criterion for its categorization. In this context, Schütte (1998, p. 71) proposed the following characteristics:

- *Organizational model* and *application system model,*
- *Functional specification*, *technical specification* and *implementation,*
- *Structural model* and *behavioral model,*
- *Object model* and *meta model,*
- *Functional model*, *data model* and *process model.*

In practice, the reference model would typically be used together with an enterprise architecture framework (e.g., TOGAF) (Lankhorst 2013, p. 24). Thus, this book proposes to structure the content of reference models according to the content that is commonly used in enterprise architecture frameworks[5]:

- The *business perspective* describes the fundamental aspects of the organization from a strategic viewpoint. In this context, a typical reference model is the *Business Model Canvas* that provides a reference structure for the development of business models (Osterwalder and Pigneur 2010).
- The *process perspective* contains, for example, business processes, responsibilities, performance indicators, and information flows. A broad variety of reference models exists for processes for different industries or functions, such as ITIL (Orand 2013) and eTOM (Kelly 2003).
- The *integration perspective* structures the components of information systems and their communication. In this context, integration architectures such as service-oriented architectures (SOA) are common concepts. A typical reference model is the *Reference Model of Open Systems Interconnection* (Zimmermann 1980, p. 425).
- The *software perspective* describes artifacts relevant for the definition of software. In the field of software engineering, patterns are an accepted concept to re-use solutions for the development of complex systems (Buschmann et al. 2007), which can be understood as reference models (Winter et al. 2009).

A further criterion for the differentiation of reference models is their maturity or acceptance in the usage community. Typically, a reference model starts as a proposed solution for a certain problem domain. If it is accepted in this problem domain, it is then re-used and further developed based on the experiences during its application. Most successful reference models belong to an organization that is recognized as an accepted association in the relevant problem domain. According to the formal function or reputation of this association, the reference model could become a de jure or de facto standard. Generally, these associations define formal procedures for the change and acceptance of their reference models as well as the involvement of their members. In this case, the reference models are a consensus of their members.

The different criteria for the categorization of reference models are summarized in Fig. 3.5.

---

[5]The different perspectives are defined according to Winter and Fischer (2007, p. 8) (cf. Sect. 3.2.2).

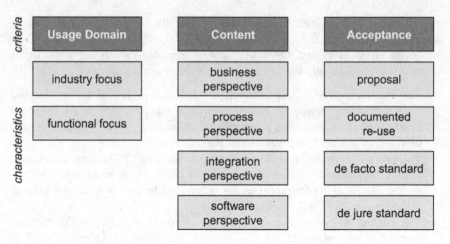

Fig. 3.5 Reference model types

3.3.3 Development and Usage of Reference Models

Reference modeling describes the development and usage of reference models (Fettke and Loos 2007a, b, p. 5; vom Brocke 2007, pp. 48–51). As for every modeling activity, reference modeling also differentiates between the model designer and the model user (Becker and Schütte 2004, p. 65). In this specific case, the model user is again a model designer who re-uses the reference model during the construction of a concrete model, which is called *application model* (vom Brocke 2007, p. 49). The challenge is that the designer of the reference model does not know its future user. Hence, the reference model is designed for an anonymous model user (Schütte 1998, p. 198; vom Brocke 2003, p. 32).

Reference modeling can be understood as a value chain (Böhmann et al. 2007, p. 129) that starts with the construction of a reference model, the usage of this reference model to construct an application model as part of a concrete solution, and the implementation of this solution as part of a concrete transformation (cf. Sect. 3.5). The real value of the reference model can only be observed after using it in a concrete implementation (Böhmann et al. 2007, p. 129). The developer of a reference model should keep the intended added value of the reference model carefully in mind.

It is advisable to consider the lessons learned regarding the reference model application during its further development. This idea is included in the iterative approach that was proposed by Schütte (1998, p. 184) (cf. Fig. 3.6):

1. The starting point is a *problem definition,* which is similar for a certain problem domain. The objective is to define a possible reference solution for exactly this problem definition. A proper understanding of the problem domain in accordance to the later model user is an indispensable prerequisite for every reference model.

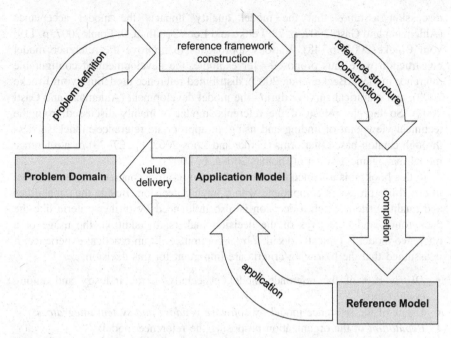

Fig. 3.6 Iterative reference modeling approach (according to Schütte 1998, p. 185)

2. The construction of a *reference model framework* is the first step from the problem to a reference solution. The framework structures the reference model. It defines the building blocks and terminology.
3. The *reference solutions* provide the content of the reference model. They are structured based on the framework and focused on the defined problem domain. The different parts of the content are defined in a hierarchical order from more high-level concepts to detailed solutions.
4. The *application* of the reference model is performed by the model user. A concrete usage scenario should have similarities to the problem definition of the reference model. In this case, the reference solutions are re-used to develop a concrete solution (the so-called application model) that should add value in the concrete context of the problem domain. Customizations are necessary according to the specific situation.
5. The *further development or extension* of the reference model should consider changes in the problem domain (e.g., technological advance, market changes) as well as lessons learned from its application. The latter requires a careful distinction between a single customization and a general requirement that should be considered in the reference model.

Through the iterative approach, the value of applying the reference model in a real-life scenario is increased. Further involvement of potential model users in the reference model construction could help to leverage its acceptance. The scientific discussion assumes that the model quality impacts the model acceptance (Ahlemann and Gastl 2007, p. 78; Fettke and Loos 2004b, p. 9; Frank 2007, p. 119; vom Brocke 2007, p. 48). Various approaches to improve the reference model construction were thus proposed—for example, the development of configurable reference models (Becker et al. 2002), distributed reference modeling (vom Brocke 2003), and empirical surveys during the model development (Ahlemann and Gastl 2007). So far, the re-use of the reference model is mainly discussed from the technical viewpoint of finding and using an appropriate reference model, such as through catalog-based platforms (Fettke and Loos 2002, pp. 27–29) or a reference model management system (Thomas 2006a, b, 2007).

In this book, it is assumed that the decision to re-use a concrete reference model in a real-life project is comparable with a buying decision. Beside the capabilities and quality criteria, such a decision is also influenced by further criteria like the background and objectives of the decision makers. In addition, the usage of a reference model is typically decided by a committee. From practical experience, it is assumed that the following criteria are important for this decision:

- *Distribution* of the reference model, especially in the industry and among competitors;
- Usage of the reference model by *software vendors and system integrators*;
- *Reputation* of the organization proposing the reference model;
- Case examples of *real-life usage projects*; and
- Availability of *trainings and support*.

The value of the reference model starts with the usage during the design of an application model. Major benefits are higher quality and efficiency due to the re-use of existing solutions (Fettke and Loos 2007a, p. 5). As reference models are generic, a customization based on specific requirements is necessary. However, a commonly accepted approach for this customization does not exist (Fettke and Loos 2007a, p. 13; Thomas 2007, p. 290).

In this context, an own analysis of 184 real-life reference model usages in the telecommunications industry provides a comprehensive empirical basis (Czarnecki 2013; Czarnecki et al. 2011, 2012). This study shows that, in most cases, a selected subset of the reference models was used. A complete implementation of the reference models was not observed in any of the 184 real-life projects. This shows that a selective usage of relevant parts of a reference model is a common approach in a practical context. Hence, completeness of usage cannot be seen as quality criteria. Moreover, it can be assumed that a successful reference model provides a comprehensive variety of reference solutions. The selection of the relevant parts is the first step of the reference model usage. The customization of those relevant reference model parts leads to an application model that is based on the reference model and considers specific requirements. On the other hand, it cannot be assumed that the reference model covers all relevant parts of the future solution. Furthermore, the future solution—which is called *solution model* in this book—might require further development that is not supported by the reference model. For this reason, there is a

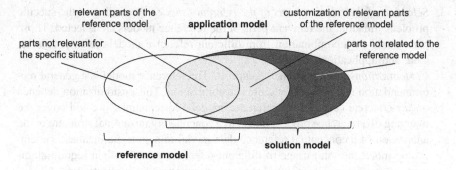

Fig. 3.7 Interrelation between reference, application, and solution model

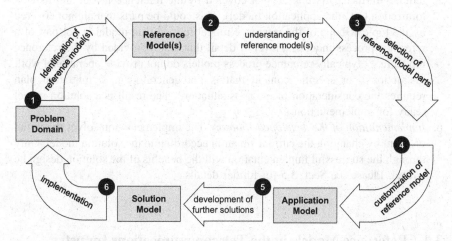

Fig. 3.8 Reference model usage

subset of content of the reference model that is relevant for the application model. This part is changed or extended due to specific customization requirements. The resulting application model is then subset of the solution model. This interrelation is illustrated in Fig. 3.7.

Accordingly, the following steps of the reference model usage are proposed (cf. Fig. 3.8):

1. *Identification of appropriate reference model(s)*: Based on the concrete problem situation, possible reference models are identified. Those could be either specific for the industry or function. Depending on the content covered, reference models could either complement one another or compete with each other.
2. *Detailed understanding of the reference model(s)*: Most reference models are complex with an own terminology and specific assumptions. Furthermore, in a practical context various stakeholders with different backgrounds and roles work together in teams. It is advisable to consider appropriate time and effort for trainings and communication of the relevant reference models.

3. *Selection of the relevant parts of the reference model*: According to the specific problem situation, the relevant parts of the reference models are selected. Those parts could be a combination from different reference models (e.g., processes, data, and applications).
4. *Customization of the reference solutions*: The reference model is a general recommendation that typically requires customization. This customization depends on the concrete scope of the reference model—for example, it could cover the mapping of a reference process model to a concrete organizational structure or the adaptation of a conceptual reference data model to a concrete database system. Furthermore, the joint usage of different reference models might require alignment of interfaces. The result of the customization is the application model.
5. *Development of further solutions*: In most cases, the concrete solution design requires further parts that are not covered by the reference model and its customization (i.e., the application model). This could be parts that are not covered by the topical scope of the reference model (e.g., a data model is not part of a reference process model), a level of detail that is not provided by the reference model (e.g., typically reference process models do not provide operational work instructions), or specific content that is not general (e.g., a migration plan requires the consideration of the as-is situation). The result is a solution model ready for implementation.
6. *Implementation of the developed solution*: The implementation solves the initial problem by changing the current situation according to the solution model. Only through the successful implementation will the benefits of the solution design be gained. Please see Sect. 3.5 for further details.

3.4   Reference Models in the Telecommunications Industry

In the telecommunications industry, the *TM Forum* is a well-known and accepted international non-profit organization. It provides a platform for the exchange of experiences and solutions with focus on the telecommunications industry.[6] The TM Forum was founded in 1988 and has more than 900 member companies. According to their own statement, the TM Forum members are representative of most organizations involved in the telecommunications industry worldwide. Members range from communication service providers to software vendors and system integrators. The TM Forum organizes regular conferences, workshops, and trainings. Furthermore, it publishes standards and best practices for the telecommunications industry. Recently the TM Forum has started to further expand to other industries, such as healthcare. In this book, the focus is on their relevant activities within the telecommunications industry.

---

[6]For general information about the TM Forum and their reference model Frameworx please see their webpage www.tmforum.org.

In this context, *Frameworx* is their term for a collection of methods, concepts, and reference models to support the transformational needs of telecommunications operators. First, the TM Forum has published those concepts under the term *NGOSS* (Kelly 2003, p. 109; Misra 2004, p. 143; Reilly and Creaner 2005), which they then changed to *Solution Framework*, and later renamed *Frameworx*. Except to the continuous update through newer versions, all three terms can be understood synonymously. In this book, the latest term *Frameworx* is used.

Frameworx consists of the following four areas (TM Forum 2015c):

1. *Business Process Framework*, also called enhanced Telecom Operations Map (eTOM), provides a hierarchical structure for business processes (TM Forum 2015a).
2. *Information Framework*, also called Shared Information/Data Model (SID), provides a structure and entity relationship model (ERM) for data (TM Forum 2015b).
3. *Application Framework*, also called Telecom Application Map (TAM), provides a hierarchical structure of functionalities for applications (TM Forum 2015d).
4. *Integration Framework*, also called Technology Neutral Architecture (TNA), provides concepts for interoperability of different systems and services (TM Forum 2012a).

The content is developed and continuously updated in working groups belonging to the TM Forum. It is based on contributions from and cooperation with member companies. Hence, it reflects a consensus of telecommunications operators, software vendors, and system integrators. All published documents are identified by a unique number (e.g., GB921-E), the version (e.g., 15.0.0) and the deliverable status (e.g., TM Forum approved). Furthermore, the International Telecommunication Union (ITU) has accepted parts of Frameworx as an official recommendation (ITU 2007, 2008a).

From a terminological perspective, the terms *framework*, *model*, *map*, and *architecture* are not clearly differentiated in the content provided by the TM Forum documents. Nevertheless, the content provided by the TM Forum is widely accepted in the telecommunications industry and has became a de facto standard. Also, our own analysis shows an intensive usage of the content in 184 real-life projects (Czarnecki 2013; Czarnecki et al. 2011, 2012). From a scientific perspective, eTOM, SID, and TAM are reference models for processes, data, and a functional application view. TNA is a collection of concepts based on a service-oriented architecture (SOA).

In addition, the IT Infrastructure Library (ITIL) is a reference model that is widely accepted in the telecommunications industry (TM Forum 2012b). ITIL was developed in the 1980' by the Central Computing and Telecommunications Agency (CCTA). It became a de facto standard for IT service management by providing a collection of best practices (Orand 2013). ITIL is independent of the telecommunications industry and is a functional reference model for the management of IT services. For telecommunications operators, it is mainly used in the context of IT organizations. So far, the ITIL 2011 Edition is the latest published version (Orand 2013).

In the following, a more detailed description of the above reference models is given. Please note that the usage of the reference models eTOM, SID, TAM, and ITIL might be subject to licensing agreements. The citations of these reference models provided in this book are for scientific and educational purpose only.

3.4.1 TM Forum Business Process Framework (eTOM)

The *Business Process Framework*, also called *enhanced Telecom Operations Map* (eTOM), is a reference process model and part of TM Forum Frameworx. It is a de facto standard in the telecommunications industry. It was accepted by the ITU in recommendation M.3050 (ITU 2007a, b, c).

The eTOM publication (version 15.0.0) is structured into the following items (TM Forum 2015c, pp. 12–13):

1. *Process definitions* are the core of the eTOM standard. This part provides a hierarchical definition of processes and sub-processes. It is structured into two documents. *Process Decompositions* (GB921-D) describes the business processes up-to level 3, while *Extended Process Decompositions* (GB921-DX) provides further details on level 4.
2. *How to guides* are a collection of various guidelines about usage and implementation of eTOM. It contains a getting started package with a primer (GB921-P), user guidelines (GB921-U and GB921-G), and frequently asked questions (GB921-Q). Furthermore, guidelines about implementing eTOM in process flows (GB921-E, GB921-F, GB921-J), and guidelines about using eTOM and ITIL (GB921-W, GB921-L, TR143) are provided.
3. *Reference material* contains release notes (RN332), construction guidelines for process flows (GB921-K), and an overview of concepts and principles (GB921-CP).
4. *Exploratory reports* provide application notes about security management (GB921-Z) and spectrum management (GB921-Y), as well as a document describing the delivery of digital services with eTOM and ITIL (TR206).
5. *eTOM models* are provided in different formats to be imported in process modeling tools as well as an MS Excel illustration, and a browsable HTML illustration.

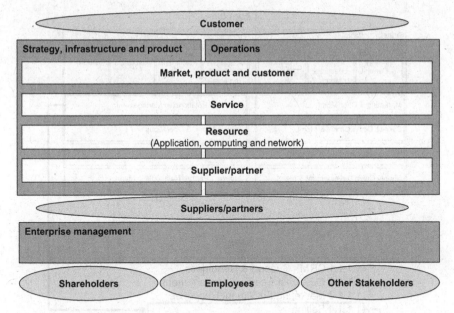

Fig. 3.9 Structure of the reference process model eTOM (Kelly 2003, p. 112)[7]

The core element of eTOM is the process definition which provides a categorization and hierarchical structure of business processes specific for telecommunications operators (Kelly 2003, pp. 110–118). On the highest level, eTOM
distinguishes the following three process groups (cf. Fig. 3.9) (Kelly 2003,
pp. 113–118):

1. *Operations* covers all processes to run a telecommunications operator with
   existing infrastructure and products, i.e., sales, after-sales, incidents, and billing.
2. *Strategy, Infrastructure and Products (SIP)* contains all other processes that are
   a prerequisite for operating a telecommunications operator—i.e., planning and
   building its infrastructure and products from strategy development to technical
   realization.

---

[7]eTOM is continuously updated and published by the TM Forum (e.g. TM Forum 2015a). In
addition, eTOM is described in various secondary publications (e.g. Czarnecki et al. 2013; Kelly
2003; Kwak et al. 2008; Misra 2004; Raouyane et al. 2011; Sathyan 2010; Yari and Fesharaki
2007). In this book, the illustration and summary of eTOM follows mainly Kelly (2003) because
the author of this publication has led the eTOM working group for several years. Please note that
the recent update of eTOM with release 15.0.0 has led to a split of the horizontal process group
Market, Product and Customer into three separate process groups (TM Forum 2015a, p. 11).

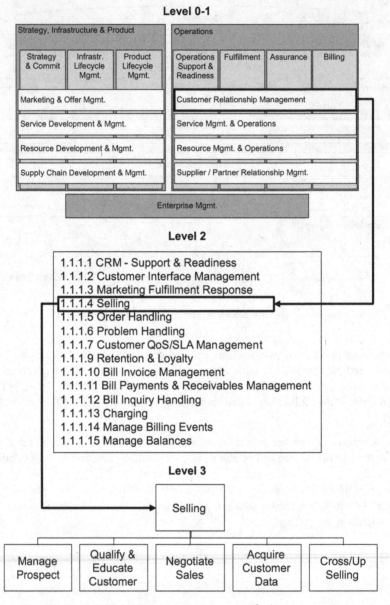

Fig. 3.10 eTOM process levels (Czarnecki et al. 2013, p. 86)[8]

3. *Enterprise Management* covers all supporting processes which are not directly involved in the core value creation—e.g., human resource management, finance, and communication.

---

[8]The interrelation between the different levels is an own illustration that was published in Czarnecki et al. (2013, p. 86). The exemplary content of the level 0–3 processes is based on eTOM (Kelly 2003, pp. 113–118). It does not include the latest changes of the horizontal process groups in release 15.0.0 which are not relevant for a general understanding of the hierarchical structure.

The core value creation in the process groups SIP and Operations is further horizontally structured by the entities involved in those processes (Kelly 2003, pp. 113–118):

- *Market, product and customer* provides the interface between the telecommunications operator and the customer.
- *Services* are an internal view on logical capabilities that are relevant to deliver products.
- *Resources* are the realization of services from a physical and logical production perspective.
- *Suppliers/partners* provides capabilities with respect to products, services, or resources.

In previous releases of eTOM, the horizontal layer for market, product, and customer was summarized in one process group (Kelly 2003, p. 112). With the recently published eTOM release 15.0.0, the entities are split into separate groups according to the domains of the reference data model SID (cf. Sect. 3.4.2) (TM Forum 2015a, p. 11).

The process definition of eTOM is structured according to those three process groups and four horizontal entities. The process decomposition provides process descriptions in a hierarchical manner (cf. Fig. 3.10) (Kelly 2003, pp. 113–118). In the following, an exemplary description of this decomposition is given. On level 0, the process group *Operations* contains the sub-group *Customer Relationship Management* which is considered as level 1. This subgroup is divided into different processes, e.g., *Selling* and *Order Handling* (both level 2). Those processes are decomposed to activities which are level 3. In this example, the level 2 process *Selling* contains the activities *Manage Prospect, Qualify and Educate Customer, Negotiate Sales, Acquire Customer Data*, and *Cross-/Up-Selling*.

eTOM is a hierarchical collection of business processes including their definition. It is helpful for a common terminology as well as a starting point for the design of business processes. Nevertheless, the original process definitions provided in the eTOM publications (i.e., GB921-D and GB921-DX) do not provide any reference for a process flow (Czarnecki et al. 2013, p. 84). For the application of a business process, the interrelation between the different process steps—i.e., the control aspect (Axenath et al. 2005, p. 48)—is crucial, and is typically provided in a graphical representation (e.g., swim lane diagram). The development of process flows based on the eTOM process definitions requires in the first step the selection of the relevant processes in the eTOM hierarchy. The second step is then to arrange them in the right order according to the logic of the process execution. In order to support these two steps (cf. Fig. 3.11), reference process flows were developed as an extension of eTOM (Czarnecki et al. 2013, p. 84).[9] They are published as an official eTOM guideline in GB921-E.

---

[9]The development of reference process flows and process domains was initiated and led by the consulting company Detecon. In their role as Managing Consultants for this company, both authors of this book have accompanied this development. Please see TM Forum (2015e) GB921-E and Czarnecki et al. (2013) for further details.

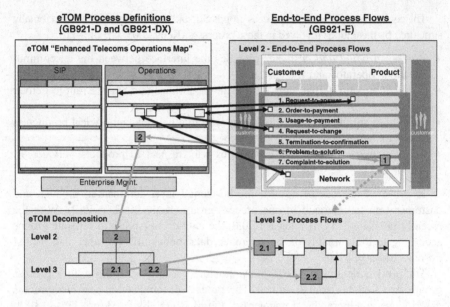

Fig. 3.11 End-to-end process flows (according to Czarnecki et al. 2013, p. 89)

The reference process flows are based on a process framework that defines end-to-end processes in different domains (Czarnecki 2013; Czarnecki et al. 2013, p. 84). Domains are a possibility in architecture development to provide a high-level business structure which is decoupled from its technical implementation (Aier and Winter 2008, pp. 180–182). In our case, the following domains are defined (Czarnecki 2013; Czarnecki et al. 2013, p. 84):

- *Customer-centric domain* covers all processes that are directly initiated by the customer.
- *Network domain* includes all processes that are related to the realization and operations of communication services and network resources.
- *Product domain* covers the whole product lifecycle from product development to product elimination.
- *Customer domain* includes all processes that deal with customers but are not directly initiated by the customer (e.g., marketing campaigns).
- *Enterprise support domain* covers all other processes that are required to run an enterprise (e.g., finance or human resource management).

For each domain end-to-end processes are defined based on use cases related to this domain. These end-to-end processes are then mapped to eTOM and further detailed by using the eTOM hierarchy (cf. Fig. 3.11). The customer-centric domain, for example, contains the end-to-end process flow *Order-to-Payment* that covers all process steps from the customer contact to the order processing and to the billing.

By combining the hierarchical process definition with end-to-end process flows, eTOM provides an extensive reference process model for telecommunications operators. Through its hierarchical structure, it can be used for strategic and planning purposes as well as for operational implementation. With the end-to-end process flows, interrelations between organizational and technical entities are transparent. The eTOM process definitions, and especially the extension through the reference process flows are used for the reference architecture described in this book (cf. Chap. 4).

3.4.2 TM Forum Information Framework (SID)

The *Information Framework*, also called *Shared Information/Data Model* (SID), is a reference data model and part of TM Forum Frameworx. Similar to eTOM, it is a de facto standard in the telecommunications industry, but in this case for data. On a high level, SID was accepted by the ITU in recommendation M.3190 (ITU 2008a).

The SID publication (version 15.0.0) is structured into the following items (TM Forum, 2015c, pp. 18–21)[10]:

1. *Standards entity definitions* provide the main content of SID. This part contains 31 separate documents (GB922…). Each of them provides a detailed definition of a data entity (e.g., Customer). This definition includes a textual definition as well as data diagrams and is structured in a hierarchical manner—i.e., each data entity is divided into different sub-entities.
2. *How to guides* are a collection of various guidelines about usage and implementation of SID. It contains a primer (GB922 Primer), an overview of the concepts and principles (GB922 Concepts and Principles) as well as two guidelines for how to use the provided data models (GB922 User's Guide and GB922-X).
3. *Model files* are provided in different formats to be imported in data modeling tools as well as an MS Excel illustration, and a browsable HTML illustration.
4. *Reference material* contains release notes (RN310) and a description of differences to previous releases (GB922 Differences).

---

[10]SID is continuously updated and published by the TM Forum (e.g. TM Forum, 2015b). In addition, SID is described in some secondary publications (e.g. Raouyane et al. 2011; Sathyan, 2010; Stamatelatos et al. 2013). This book provides a high-level summary of SID mainly based on TM Forum (2015b).

The objective of SID is to provide a reference for data objects from a business perspective. The focus is on a logical data view which is independent of any physical implementation in a database system. Similar to eTOM, SID is also structured in a hierarchical manner (Sathyan 2010, p. 381; TM Forum 2015b, p. 10):

- *Domain* is the highest level of aggregation. It provides a first structure and classification according to business areas.
- *Aggregated Business Entity* (ABE) is a collection of different business entities (i.e., data objects). The aggregation of business entities to ABEs is decided based on their topical context, e.g., the ABE *Customer* contains all business entities that are required to describe a customer.
- *Business entities* are logical data objects that are described from a business perspective. They could be tangible (e.g., customer), abstract (e.g., subscriber), or transactional (e.g., customer order).
- *Attributes* are used to further detail business entities.
- *Relations* are defined between business entities.

On the highest level, SID is structured into the following eight domains (Sathyan 2010, pp. 383–384; TM Forum 2015b, p. 9):

1. *Market/Sales domain* contains ABEs that are necessary to understand the market and to plan sales activities (e.g., market segments, sales channels).
2. *Product domain* covers ABEs that are related to the product lifecycle and range from definition of products (e.g., product specification) to product usage (e.g., product usage statistic).
3. *Customer domain* aggregates all business entities for defining, contacting, serving, and billing of customers. It includes basic ABEs (e.g., customer) as well as transactional ABEs which deal directly with the customer (e.g., customer order).
4. *Service domain* contains ABEs that are relevant for the whole service lifecycle and range from service definition (e.g., service specification) to service operations (e.g., service trouble).
5. *Resource domain* covers all ABEs that are required for the resource lifecycle, from resource definition (e.g., resource topology) to resource operations (e.g., resource trouble).
6. *Supplier/Partner domain* includes ABEs that are required in interactions with suppliers and partners from planning data (e.g., Supplier/Partner plan) to operational transactions (e.g., Supplier/Partner order).
7. *Common Business Entities domain* contains ABEs that are required to define general data objects, such as location.
8. *Enterprise domain* covers further specific ABEs relevant for the enterprise, such as enterprise security.

The eight SID domains are linked to the horizontal structure of eTOM—for example, the SID domain *Resource* is related to the horizontal eTOM process groupings *Resource Development Management* and *Resource Management and Operations*. This supports the consistency between both reference models. In a concrete application model, the data should be consistently related to the input and output of the processes. In prior releases of eTOM, the SID domains *Market/Sales*, *Product* and *Customer* were summarized in a single horizontal process group. With release 15.0.0, the horizontal structure of eTOM is now completely based on the SID domains (TM Forum 2015c, p. 10).

The SID domains are then further detailed in ABEs. Each ABE consists of various business entities which have relations to other business entities. The further detailing of SID domains and ABEs is explained in the following example (TM Forum 2011, p. 23): The *Customer* domain includes, among others, the ABEs *Customer* and *Customer Order*. The ABE *Customer* is detailed in several business entities, e.g., *Customer* and *Customer Account*. Following the same logic, the ABE *Customer Order* is detailed in several business entities, e.g., *Customer Order* and *Customer Order Item*. There is a relation between the business entities *Customer* and *Customer Account*, *Customer Account* and *Customer Order*, and *Customer Order* and *Customer Order Item*.

The SID standard entity definitions provide detailed descriptions of each SID domain, ABE, and business entity. The relations between business entities are defined in Entity Relationship Models (ERM) following the notation of the Unified Modeling Language (UML).

Based on the relations between the business entities, relations between ABEs can also be derived. The understanding of the relation between major SID ABEs is a helpful background for applying the TM Forum reference models (TM Forum 2011, p. 23): *Customers* belong to a *market* and buy *products*. Products are composed of *services* and realized by *resources*. Buying a product leads to a *customer order*, which is a *business transaction*. The *customer order* includes the *product(s)* which are delivered through *services*. The *services* that are directly relevant for products are called *customer-facing services*. *Services* that are mapped to *resources* are called *resource-facing services*. *Resources* are divided into *logical* and *physical resources*. A *customer* who uses a *service* is called *user*. A *customer* and *resources* belong to a *location*. *Products*, *services*, and *resources* may be provided by *supplier and partners*. The development and operations of *resources* is *work* that might be supported by *supplier and partners*.

SID provides a comprehensive reference for a logical data model. It ranges from a high-level structure to detailed definitions of business entities and their relations. The general understanding of major business entities in combination with eTOM is an especially important basis for telecommunications operators. Therefore, it is used for the reference architecture described in this book (cf. Chap. 4).

3.4.3   TM Forum Application Framework (TAM)

The *Application Framework*, also called *Telecom Application Map* (TAM), is a reference function model and part of TM Forum. The objective of TAM is to provide a structure of functions from an applications perspective (TM Forum 2015d, p. 6). In this case, applications are a logical definition and clustering of functionalities that could be realized through concrete software products. The requirements for those applications are derived from eTOM and SID. The TM Forum understands *functionalities* as a verbal definition of capabilities that are required to fulfill a concrete task (TM Forum 2015d, p. 9).

The TAM publication (version 15.0.0) is structured into the following items (TM Forum 2015c, p. 26)[11]:

1. *Standards application definitions* provide the main content of TAM. It contains one document (GB929-D) with a textual definition of functionalities. Similar to eTOM, the functionalities are structured in a hierarchical manner.
2. *How to guide* is a document (GB929-U) that describes different usage scenarios of TAM.
3. *Model files* are provided in different formats, e.g., as an MS Excel illustration, and a browsable HTML illustration.
4. *Functional decomposition* provides a detailed list (GB929-F) related to TAM that can be used to define system requirements for a service provider.
5. *Reference material* contains a release note (RN315) and an explanation of TAM concepts and principles (GB929-CP).

TAM defines an application by clustering different functionalities. Those applications are again grouped in order to provide a hierarchical structure. On the highest level, TAM is structured into horizontal domains which are exactly the SID domains (TM Forum 2015f, p. 20):

1. *Market/Sales domain* contains functionalities that are required to understand the market as well as to plan and coordinate sales activities (e.g., Sales Account Management).
2. *Product domain* covers the functionalities of the whole product lifecycle from strategic product propositions to operational product performance. A major functionality is the Product Catalog Management.
3. *Customer domain* deals with all functionalities that are related to the customer. This domain has a broad scope like e.g. understanding the customer (e.g., Customer Information Management), serving the customer (e.g., Customer Order Management), and billing the customer (e.g., Billing Account Management).

---

[11]TAM is continuously updated and published by the TM Forum (e.g. TM Forum 2015d). This book provides a high-level summary of TAM mainly based on TM Forum (2015d).

4. *Service domain* provides functionalities to realize and operate services that are required for the offered products. It has interdependencies to the Product and Customer domains: for example, the Service Catalog Management is the equivalent to the Product Catalog Management and the Service Order Management supports the realization of customer orders.

5. *Resource domain* covers all functionalities in order to realize and operate resources. Resources are the technical realization of services. There are inter-dependencies to the Service domain. Functionalities range from defining and itemizing resources (e.g., Resource Inventory Management) to monitoring resource operations (e.g., Resource Performance Management).

6. *Supplier/Partner domain* includes functionalities to deal with suppliers and partners as well as the management of payments with wholesale and intercon-nect partners.

7. *Enterprise domain* covers functionalities that are required for the general management of a telecommunications operator (e.g., Financial Management).

The seven domains above are directly derived from SID. In addition, TAM includes the following two specific domains (TM Forum 2015f, p. 20):

8. *Integration Infrastructure domain* contains functionalities that are required to define and realize the integration of different applications and their respective software systems (e.g., Enterprise Application Integration).

9. *Cross domain* includes functionalities that are relevant for different domains (e.g., Catalog Management).

In addition to the vertical domains, TAM follows the vertical structure from eTOM. The functionalities from the first seven TAM domains are linked to the following eTOM process groups: (1) Strategy, Infrastructure and Product, (2) Operations, Support and Readiness, (3) Fulfillment, (4) Assurance, and (5) Billing.

On the detailed levels of TAM, functionalities and applications are grouped according to the following criteria (TM Forum 2015d, pp. 11–12):

- *Invocation context* combines applications that are using the same functionalities.
- From the *end user* perspective, applications are grouped that are required by the same end user for his daily work.
- A similar *purpose* is seen as a commonality which leads to a grouping of applications.

TAM provides a hierarchical structure of functionalities which considers busi-ness requirements defined by eTOM and SID. The hierarchical structure is based on different levels of detail (cf. Fig. 3.12), which is comparable with the process levels used by eTOM. In general, TAM ranges from level 0 domains to level 4 applica-tions. However, not all parts of TAM are decomposed to level 4. The structure of TAM is explained based on the following example (TM Forum 2015f, pp. 22, 39). On level 0, TAM contains the different domains described above. These domains are further decomposed to applications which are the level 1—e.g., the *Market & Sales* domain contains the applications *Sales Portals* and *Contract Management,* among others. Those applications are further detailed on level 2—e.g., *Sales*

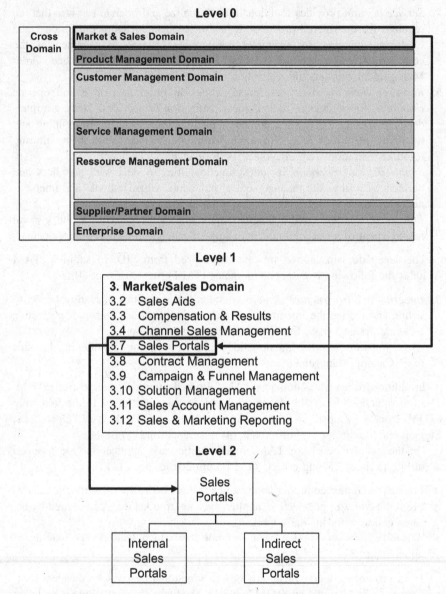

Fig. 3.12 Hierarchical structure of TAM (Czarnecki 2013, p. 60)[12]

Portals are differentiated into *Internal Sales Portals* and *Indirect Sales Portals*. In addition, TAM provides a textual description of the functionalities that belong to each application.

---

[12]The interrelation between the different levels is an own illustration that was published in Czarnecki (2013, p. 60). The exemplary content of the level 0–2 functionalities is based on TAM (TM Forum 2015f, pp. 22, 39).

The TAM structure and definitions are a reference for functions from an applications perspective. It is comparable with a functional tree, which is a common tool to define functional requirements. Together, eTOM, SID, and TAM are used as the basis for the reference architecture explained in Chap. 4.

3.4.4   IT Infrastructure Library (ITIL)

The *IT Infrastructure Library* (ITIL) was developed in the 1980s by the *Central Computing and Telecommunications Agency* (CCTA). It provides a collection of best practices for the management and provisioning of IT services and became a de facto standard for IT service management. In 2007, ITIL version 3 was published, which was updated in 2011 (ITIL 2011 Edition). It is structured into the following core publications (Long 2012, pp. 3–95):

- *Service Strategy* covers all strategic aspects of IT service management. It ranges from the definition of the service portfolio and its financial management to the management of business relations.
- *Service Design* contains all activities necessary for coordinating, planning, and conducting the development of IT services. It also includes the management of a service catalog, service level agreements, and the management of suppliers.
- *Service Transition* deals with the roll-out of designed services including transition planning, change management, configuration management, and testing.
- *Service Operation* covers typical operation tasks, such as the management of problems, incidents, and requests.
- *Continual Service Improvement* includes a seven-step improvement process as well as the measurement and reporting of services.

ITIL provides detailed recommendations for the above topics. It is a comprehensive reference model for an IT organization dealing with IT service management. The scope of ITIL is limited from a functional perspective but not specific for any industry. In the telecommunications industry, ITIL is commonly accepted for IT service management and can be found in various IT departments of service providers around the world.

In comparison to ITIL, the reference process model eTOM (cf. Sect. 3.4.1) is a de facto standard for all business processes of service providers and specific for the telecommunications industry. Therefore, the TM Forum and the IT Service Management Forum (itSMF) published a joint view on how to combine eTOM and ITIL (TM Forum 2012b). From their perspective, eTOM and ITIL are offering complementary content. ITIL is focused on IT practices and should be linked to the business environment and business processes. eTOM provides a hierarchical perspective on business processes that require a link to underlying business support tasks (TM Forum 2012b, p. 10). Therefore, they assume that linking eTOM with ITIL would be of mutual value (TM Forum 2012b, p. 16).

However, it is important to understand the exact scope of both reference models in order to define their interrelations. One essential point is that both eTOM and ITIL use the term *service* in a different meaning. In eTOM the term *service* is understood as a communication service. A fundamental idea of the TM Forum is that communication products are divided into communication services which are realized by resources. This implies that communication services are an internal construct that is not seen by customers and irrespective of its technical realization. This structure of communication products, services, and resources is specific for the telecommunications industry. It is an integral part of the structure and concepts of all three TM Forum reference models (eTOM, SID and TAM). In contrast, ITIL understands the term *service* as IT service that is provided by IT organizations and systems. In ITIL, a service can be used within an enterprise to support business areas or external customers/users.

IT services (ITIL) can be involved in communication services or resources (eTOM). The decoupling of a communication service from its technical realization (resource) is fundamental for understanding telecommunication concepts such as Next Generation Networks (NGN) . This idea is not reflected in IT services as they are provided in ITIL. Therefore, ITIL cannot simply be mapped to the service layer of eTOM (TM Forum 2012b, p. 15). Moreover, ITIL is related to various business processes and can be used to specify and improve them on an operational level.

3.5   Introduction to Enterprise Transformation

Enterprises are continuously confronted with changes of market requirements, customer demands, and technological conditions. From an economic perspective, the ability to continuously adapt to technical innovations is an important driver for sustainable wealth (Hanna 2010, pp. 28–33). This leads to the permanent requirement of enterprises to react to these changed conditions (Rouse 2006a, p. 1). The continuous adaptation of an enterprise to internal and external forces is a key of entrepreneurship. It is not a new topic and has been discussed under various headlines in the past, including *rightsizing*, *business process reengineering*, and *lean thinking* (Rouse 2006a, pp. 1–4).

The term *enterprise transformation* can be understood as systematic change of an enterprise from an initial state to a target state (Alt and Zerndt 2009, p. 48; Lahrmann et al. 2012, p. 255). However, there is a difference between a minor routine change and a fundamental change (Rouse 2005, pp. 279–280). Most authors understand a transformation as a fundamental change of an enterprise's value proposition, structures, and/or technologies (Hanna 2010, p. 15; Lahrmann et al. 2012, p. 253; Rouse 2005, pp. 279–280). A routine change is part of the daily business and could be initiated by a continuous process improvement (Rouse 2005, p. 279). Typical transformations could include an organizational restructuring, the implementation of a new IT system along with optimized processes, or the out-sourcing of a support function. Some of the general drivers for transformations are new technologies, globalized markets, or mergers and acquisitions (Aier and Weiss 2012, p. 1073). As a transformation is a fundamental change of an enterprise, its realization is a complex and challenging task (Rouse 2006a, p. 1).

From the information systems perspective, enterprise transformation deals with the dynamic aspect of implementing the designed solution (e.g. Aier and Gleichauf 2010a; Aier and Weiss 2012; Alt and Zerndt 2009; Jetter et al. 2009; Young and Johnston 2003). The solution design can be seen as part of a value chain that ends with the transformation in order to create a positive impact for the enterprise (Böhmann et al. 2007, p. 129). A study of Lahrmann et al. (2012, p. 259) shows that most enterprise transformations aim to achieve cost reduction and revenue increase through business optimization, changed operating models, or standardized pro-cesses and platforms.

Describing a transformation means a transition path from an as-is to a target state with a well-defined number of transition states (Aier and Gleichauf 2010a). A transformation can be described by modeling the different states, e.g., through the application of an enterprise architecture framework (Aier and Gleichauf 2010a). The as-is enterprise architecture is documented as starting point. The target enter-prise architecture is defined as the goal of the enterprise transformation. The way from the as-is to the target architecture is described in the transition model, which defines the changes from one state to the other framework (Aier and Gleichauf 2010a) (cf. Fig. 3.13).

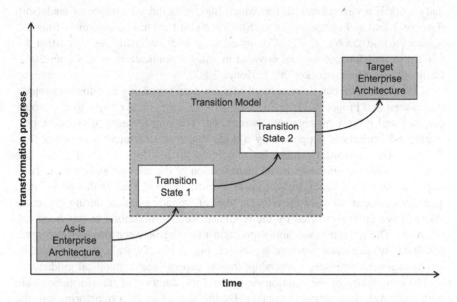

Fig. 3.13 Transformation from as-is to target architecture

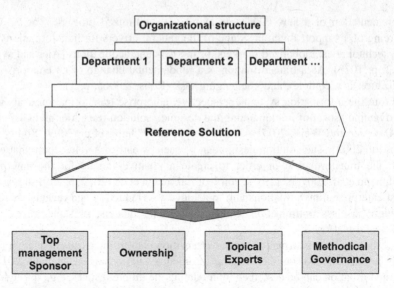

Fig. 3.14 Identification of relevant stakeholders based on reference solution

Certainly a successful enterprise transformation requires an appropriate modeling of the targeted solution and the transition path. An enterprise transformation is an interdisciplinary endeavor that necessitates aligned changes in various parts of an enterprise, such as strategy, technology, skills, offerings, and organization (Rouse 2005, p. 286). Furthermore, the challenge of convincing people to change their daily work is a key success factor, which highlights the importance of leadership (George 2006) and cultural change (Shields 2006) for optimal enterprise transformation (Rouse 2006b, p. 27). Communication and leadership are important for successful transformations, and covered in various publications (e.g. Carter 2013; Collins 2001; Koenigsaecker 2013; Kotter 2007).

A transformation can be structured as a program consisting of different projects (Greefhorst and Proper 2011, pp. 13–14). From a methodical perspective, a proper program and project management is required. The management of complex programs and projects is supported by a wide range of structured approaches (e.g. Brown 2008; Harvard Business Review Press 2013; Thiry 2010). Besides this general methodical guidance, a fast identification of the relevant stakeholders is an important step for a successful set-up of a transformation. One challenge is that a high-level understanding of the relevant topics is already essential during the set-up stage. This step is supported by the reference solution described in this book (cf. Chap. 4). The reference solution supports a mapping between the concrete organizational structure and required topics (cf. Fig. 3.14). Typical categories can be top-management sponsors, ownership, topical experts, and methodical guidance.

The complexity of the solution design and the duration of transformations are both challenges in a practical context. During the set-up of a transformation, the commitment and motivation is typically high. The solution design requires time and resources for detailed work, the consideration of interrelations, and continuous decisions. There is a risk that priorities change before the transformation is implemented. An accompanying change management approach (e.g. Kotter 2012, pp. 37–168) is recommended. Also, the re-usage of proven reference solutions helps to accelerate the solution design (cf. Chap. 4). Specific success factors for planning and implementing transformations in the telecommunications industry are discussed in Chap. 5. From a generic perspective, they can be summarized as follows:

- *Set-up of a program and project structure* that considers the different strategic, tactical, and operational aspects of the enterprise transformation (Greefhorst and Proper 2011, pp. 13–14).
- *Identification and involvement of the relevant stakeholders* from top management to operational level. Long-term commitment to support the whole transformation is essential.
- *Continuous communication of importance and target picture* from the top-management to the whole organization (Kotter 2012, p. 16).
- *Involvement of responsible personnel from the day-to-day business* in solution design in order to support a successful handover to execution.
- *The right balance between the hierarchical levels* is essential, especially for steering, communication, decisions, and approvals.
- *Empowerment of project organization* for a fast decision making is essential. The complexity of the solution design requires continuous decisions and approvals.

References

Ahlemann, F. (Ed.). (2012). *Strategic enterprise architecture management: challenges, best practices, and future developments, Management for professionals.* New York: Springer, Berlin.

Ahlemann, F., & Gastl, H. (2007). Process model for an empirically grounded reference model construction. In P. Fettke & P. Loos (Eds.), *Reference Modeling for Business Systems Analysis* (pp. 77–97). Hershey, PA: Idea Group Publishing.

Aier, S., & Gleichauf, B. (2010a). Towards a Systematic approach for capturing dynamic transformation in enterprise models. In *Forty-Third Annual Hawaii International Conference on System Sciences (HICSS-43)* (pp. 1–10). IEEE Computer Society.

Aier, S., & Gleichauf, B. (2010b). Applying design research artifacts for building design research artifacts: A process model for enterprise architecture planning. In R. Winter, J. L. Zhao, & S. Aier (Eds.), *Global Perspectives on Design Science Research* (pp. 333–348). Berlin, Heidelberg: Springer.

Aier, S., Gleichauf, B., & Winter, R. (2011). Understanding enterprise architecture management design—An empirical analysis. In A. Bernstein, G. Schwabe (Eds.), *Proceedings of the 10th International Conference on Wirtschaftsinformatik* (pp. 645–654). Zürich.

Aier, S., Kurpjuweit, S., Saat, J., & Winter, R. (2009). Enterprise architecture design as an engineering discipline. *AIS Transactions on Enterprise Systems, 1,* 36–43.

Aier, S., Kurpjuweit, S., Schmitz, O., Schulz, J., Thomas, A., & Winter, R. (2008). An engineering approach to enterprise architecture design and its application at a financial service provider. *Modellierung Betrieblicher Informationssysteme (MobIS 2008)* (pp. 115–130). Saarbrücken: GI/Köllen.

Aier, S., & Weiss, S. (2012). Facilitating enterprise transformation through legitimacy—An institutional perspective. In D. C. Mattfeld & S. Robra-Bissantz (Eds.), *Multi-Konferenz Wirtschaftsinformatik 2012 - Tagungsband Der MKWI 2012* (pp. 1073–1084). Braunschweig: GITO Verlag.

Aier, S., & Winter, R. (2008). Virtuelle Entkopplung von fachlichen und IT-Strukturen für das IT/Business Alignment—Grundlagen, Architekturgestaltung und Umsetzung am Beispiel der Domänenbildung. *Wirtschaftsinformatik, 51,* 175–191. doi:10.1007/s11576-008-0115-0

Alpar, P., Alt, R., Bensberg, F., Grob, H. L., Weimann, P., & Winter, R. (2014). *Anwendungsorientierte Wirtschaftsinformatik: strategische Planung, Entwicklung und Nutzung von Informationssystemen, 7, aktualisierte und* (Aufl ed.). Springer Vieweg, Wiesbaden: Lehrbuch.

Alt, R., & Zerndt, T. (2009). Grundlagen der transformation. In R. Alt, B. Bernet, & T. Zerndt (Eds.), *Transformation von Banken Praxis des In- und Outsourcings in der Finanzindustrie* (pp. 47–68). Berlin: Springer.

Avison, D., & Fitzgerald, G. (2006). *Information systems development: Methodologies, techniques & tools* (4th ed.). London: McGraw-Hill.

Axenath, B., Kindler, E., & Rubin, V. (2005). An open and formalism independent meta-model for business processes. In E. Kindler & M. Nüttgens (Eds.), *Proceedings of the Workshop on Business Process Reference Models 2005 (BPRM 2005)* (pp. 45–59). France: Nancy.

Becker, J., & Delfmann, P. (2007). *Reference modeling efficient information systems design through reuse of information models.* Heidelberg: Physica Verlag.

Becker, J., Delfmann, P., & Knackstedt, R. (2002). Eine Modellierungstechnik für die konfigurative Referenzmodellierung. In J. Becker, H. L. Grob, S. Klein, Kuchen, H., Müller-Funk, U., Vossen, G. (Eds.), *Referenzmodellierung 2002. Methoden - Modelle - Erfahrungen, Arbeitsberichte Des Instituts Für Wirtschaftsinformatik* (pp. 35–79). Münster.

Becker, J., Kugeler, M., & Rosemann, M. (2003). *Process management a guide for the design of business processes.* Berlin: Springer.

Becker, J., & Schütte, R. (2004). Handelsinformationssysteme, Redline Wirtschaft. mi-Wirtschaftsbuch.

Böhmann, T., Schermann, M., & Krcmar, H. (2007). Application-oriented evaluation of the SDM reference model: Framework, instantiation and initial findings. In J. Becker & P. Delfmann (Eds.), *Reference Modeling* (pp. 123–144). Heidelberg: Physica-Verlag.

Bolstorff, P., & Rosenbaum, R. G. (2012). *Supply chain excellence: A handbook for dramatic improvement using the SCOR model,* 3rd ed. New York, NY: AMACOM, American Management Association.

Brown, J. T. (2008). *The handbook of program management: How to facilitate project success with optimal program management.* New York: McGraw-Hill.

Buckl, S., Ernst, A. M., Matthes, F., Ramacher, R., & Schweda, C. M. (2009). Using enterprise architecture management patterns to complement TOGAF. In *Proceedings of the 13th IEEE International Enterprise Distributed Object Computing Conference (EDOC 2009)* (pp. 34–41). Auckland, New Zealand: IEEE. doi:10.1109/EDOC.2009.30

Buschmann, F., Henney, K., & Schmidt, D. C. (2007). *Pattern oriented software architecture volume 5: On patterns and pattern languages.* Chichester [u.a.]: Wiley.

Carter, L. (Ed.). (2013). *The change champion's field guide: Strategies and tools for leading change in your organization,* 2nd edn., and updated. San Francisco. Calif Wiley.

Collins, J. (2001). *Good to great: Why some companies make the leap... and others don't.* London: Random House.

Czarnecki, C. (2013). *Entwicklung einer referenzmodellbasierten Unternehmensarchitektur für die Telekommunikationsindustrie.* Berlin: Logos-Verl.

Czarnecki, C., Winkelmann, A., & Spiliopoulou, M. (2011). Making business systems in the telecommunication industry more customer-oriented. In J. Pokorny, V. Repa, K. Richta, W. Wojtkowski, H. Linger, C. Barry, & M. Lang (Eds.), *Information Systems Development* (pp. 169–180). New York: Springer.

Czarnecki, C., Winkelmann, A., & Spiliopoulou, M. (2012). Transformation in telecommunication—Analyse und clustering von real-life projekten. In D. C. Mattfeld & S. Robra-Bissantz (Eds.), *Multi-Konferenz Wirtschaftsinformatik 2012—Tagungsband Der MKWI 2012* (pp. 985–998). Braunschweig: GITO Verlag.

Czarnecki, C., Winkelmann, A., & Spiliopoulou, M. (2013). Reference process flows for telecommunication companies: An extension of the eTOM model. *Business & Information Systems Engineering, 5*, 83–96. doi:10.1007/s12599-013-0250-z

Desfray, P., & Raymond, G. (2014). Modeling enterprise architecture with TOGAF: A practical guide using UML and BPMN.

Engels, G., Hausmann, J. H., Heckel, R., & Sauer, S. (2000). Dynamic meta modeling: A graphical approach to the operational semantics of behavioral diagrams in UML. In A. Evans, S. Kent, & B. Selic (Eds.), *≪UML≫ 2000—The Unified Modeling Language* (pp. 323–337). Berlin Heidelberg, Berlin, Heidelberg: Springer.

Ferstl, O. K., & Sinz, E. J. (2008). Grundlagen der Wirtschaftsinformatik, 6., überarb. und Aufl. ed. Oldenbourg, München.

Fettke, P., & Loos, P. (2002). Methoden zur Wiederverwendung von Referenzmodellen—Übersicht und Taxonomie. In J. Becker, H. L. Grob, S. Klein, H. Kuchen, U. Müller-Funk, & G. Vossen (Eds.), *Referenzmodellierung 2002* (pp. 35–79). Münster: Methoden - Modelle - Erfahrungen, Arbeitsberichte Des Instituts Für Wirtschaftsinformatik.

Fettke, P., & Loos, P. (2004a). Referenzmodellierungsforschung. *Wirtschaftsinformatik, 46*, 331–340.

Fettke, P., & Loos, P. (2004b). Referenzmodellierungsforschung: Langfassung eines Aufsatzes (No. 16), Working Papers of the Research Group Information Systems & Management. Johannes Gutenberg-University Mainz, Mainz.

Fettke, P., & Loos, P. (2007a). Perspectives on reference modeling. In P. Fettke & P. Loos (Eds.), *Reference Modeling for Business Systems Analysis* (pp. 1–21). IGI Global.

Fettke, P., & Loos, P. (Eds.). (2007b). *Reference modeling for business systems analysis*. Hershey, PA: Idea Group Pub.

Fettke, P., Loos, P., & Zwicker, J. (2006). Business process reference models: Survey and classification. In C. J. Bussler & A. Haller (Eds.), *Business Process Management Workshops* (pp. 469–483)., Lecture Notes in Computer Science Berlin, Heidelberg: Springer.

Frank, U. (2007). Evaluation of reference models. In P. Fettke & P. Loos (Eds.), *Reference Modeling for Business Systems Analysis* (pp. 118–144). Hershey, PA: Idea Group Publishing.

George, W. (2006). Transformational leadership. In W. B. Rouse (Ed.), *Enterprise Transformation: Understanding and Enabling Fundamental Change* (pp. 69–78). Hoboken, N.J.: Wiley-Interscience.

Greefhorst, D., & Proper, E. (2011). *Architecture principles: The cornerstones of enterprise architecture*. Berlin, Heidelberg: Springer.

Hanna, N. (2010). *Enabling enterprise transformation: business and grassroots innovation for the knowledge economy, Innovation, technology, and knowledge management*. London: Springer, New York.

Harel, D., & Rumpe, B. (2004). Meaningful modeling: what's the semantics of "semantics"? *Computer, 37*, 64–72. doi:10.1109/MC.2004.172

Harvard Business Review Press. (2013). HBR's guide to project management.

IEEE Computer Society. (Ed.). (2000). *IEEE recommended practice for architectural description (IEEE Std 1471-2000)*. New York: Institute of Electrical and Electronics Engineers.

ITU. (2007a). ITU-T recommendation M.3050.1: Enhanced telecom operations map (eTOM)—The business process framework.

ITU. (Ed.). (2007b). ITU-T recommendation M.3050.0: Enhanced telecom operations map (eTOM)—Introduction.

ITU. (2007c). ITU-T recommendation M.3050.2: Enhanced telecom operations map (eTOM)—Process decompositions and descriptions.

ITU. (Ed.). (2008a). ITU-T recommendation M.3190: Shared information and data model (SID).

Jetter, M., Satzger, G., & Neus, A. (2009). Technological innovation and its impact on business model, organization and corporate culture—IBM's transformation into a globally integrated, service-oriented enterprise. *Business & Information Systems Engineering, 1,* 37–45. doi:10.1007/s12599-008-0002-7

Karagiannis, D., & Woitsch, R. (2010). Knowledge engineering in business process management. In J. vom Brocke & M. Rosemann (Eds.), *Handbook on Business Process Management 2* (pp. 463–485). Berlin, Heidelberg: Springer Berlin Heidelberg.

Kelly, M. B. (2003). The telemanagement forum's enhanced telecom operations map (eTOM). *Journal of Network and Systems Management, 11,* 109–119.

Koenigsaecker, G. (2013). *Leading the lean enterprise transformation* (2nd ed.). Boca Raton: CRC Press.

Kotter, J. P. (2007). Leading change: Why transformation efforts fail. *Harvard Business Review, 85,* 96.

Kotter, J. P. (2012). *Leading change.* Boston, Mass: Harvard Business Review Press.

Krcmar, H. (2005). *Informationsmanagement.* Berlin, u. a.: Springer.

Kühne, T. (2006). Matters of (Meta-) modeling. *Software & Systems Modeling, 5,* 369–385. doi:10.1007/s10270-006-0017-9

Kurpjuweit, S., & Winter, R. (2007). Viewpoint-based meta model engineering. In EMISA. p. 2007.

Kwak, E., Chang, B.-Y., Hong, D. W., & Chung, B. (2008). A study on the service quality management process and its realization strategy for capturing customer value. In Y. Ma, D. Choi, & S. Ata (Eds.), *Challenges for Next Generation Network Operations and Service Management* (pp. 297–306). Berlin Heidelberg, Berlin, Heidelberg: Springer.

Lahrmann, G., Labusch, N., Winter, R., & Uhl, A. (2012). Management of large-scale transformation programs: State of the practice and future potential. In S. Aier, M. Ekstedt, F. Matthes, E. Proper, & J. L. Sanz (Eds.), *Trends in Enterprise Architecture Research and Practice-Driven Research on Enterprise Transformation* (pp. 253–267). Berlin Heidelberg, Berlin, Heidelberg: Springer.

Lankhorst, M. (Ed.). (2013). *Enterprise architecture at work: modelling, communication and analysis, 3* (ed ed.). Berlin: The Enterprise Engineering Series. Springer.

Laudon, K. C., & Laudon, J. P. (2012). *Management information systems: Managing the digital firm* (12th ed.). Boston: Prentice Hall.

Long, J. O. (2012). *ITIL 2011 at a glance.* New York: Springer.

Ludewig, J. (2003). Models in software engineering? An introduction. *Software & Systems Modeling, 2,* 5–14. doi:10.1007/s10270-003-0020-3

Misra, K. (2004). *OSS for telecom networks: An introduction to network management.* London, u. a.: Springer.

Noran, O. (2003). A mapping of individual architecture frameworks (GRAI, PERA, C4ISR, CIMOSA, ZACHMAN, ARIS) onto GERAM. In P. Bernus, L. Nemes, & G. Schmidt (Eds.), *Handbook on Enterprise Architecture* (pp. 65–210). Berlin Heidelberg, Berlin, Heidelberg: Springer.

Orand, B. (2013). Foundations of IT service management: With ITIL 2011.

Ostadzadeh, S. S., Aliee, F. S., & Ostadzadeh, S. A. (2007). A method for consistent modeling of Zachman framework cells. In K. Elleithy (Ed.), *Advances and Innovations in Systems, Computing Sciences and Software Engineering* (pp. 375–380). Netherlands, Dordrecht: Springer.

Osterwalder, A., & Pigneur, Y. (2010). *Business model generation: A handbook for visionaries, game changers, and challengers.* Hoboken, NJ: Wiley.

Perks, C., & Beveridge, T. (2003). *Guide to enterprise IT architecture.* Springer, New York: Springer professional computing.

Pousttchi, K., & Hufenbach, Y. (2011). Value creation in the mobile market: A reference model for the role(s) of the future mobile network operator. *Business & Information Systems Engineering, 3,* 299–311. doi:10.1007/s12599-011-0175-3

Raouyane, B., Bellafkih, M., Errais, M., Leghroudi, D., Ranc, D., & Ramdani, M. (2011). eTOM business processes conception in NGN monitoring. In S. Lin & X. Huang (Eds.), *Advanced Research on Computer Education, Simulation and Modeling* (pp. 133–143). Berlin Heidelberg, Berlin, Heidelberg: Springer.

Reilly, J. P., & Creaner, M. J. (2005). NGOSS distilled: The essential guide to next generation telecoms management. The Lean Corporation.

Rosemann, M. (2003). Preparation of process modeling. In J. Becker, M. Kugeler, & M. Rosemann (Eds.), *Process Management* (pp. 41–78). Berlin Heidelberg, Berlin, Heidelberg: Springer.

Rouse, W. B. (2005). A theory of enterprise transformation. *Systems Engineering, 8*, 279–295. doi:10.1002/sys.20035

Rouse, William B. (2006a). Introduction & overview. In W. B. Rouse (Ed.), *Enterprise Transformation: Understanding and Enabling Fundamental Change* (pp. 1–16). Hoboken, N. J.: Wiley-Interscience.

Rouse, William B. (2006b). Enterprises as systems. In W. B. Rouse (Ed.), *Enterprise Transformation: Understanding and Enabling Fundamental Change* (pp. 17–38). Hoboken, N.J.: Wiley-Interscience.

Rouse, W. B., & Baba, M. L. (2006). Enterprise transformation. *Communications of the ACM, 49*, 66–72. doi:10.1145/1139922.1139951

Rummler, G. A., & Ramias, A. J. (2010). A framework for defining and designing the structure of work. In J. vom Brocke & M. Rosemann (Eds.), *Handbook on Business Process Management 1* (pp. 83–106). Springer Berlin Heidelberg, Berlin, Heidelberg.

Saat, J., Franke, U., Lagerstrom, R., & Ekstedt, M. (2010). Enterprise architecture meta models for IT/business alignment situations. In *Proceedings of the 14th IEEE International Enterprise Distributed Object Computing Conference (EDOC 2010)* (pp. 14–23). IEEE. doi:10.1109/EDOC.2010.17

Sathyan, J. (2010). Fundamentals of EMS, NMS, and OSS/BSS. Boca Raton, Fla: CRC Press: Auerbach Publications.

Satzinger, J. W. (2015). *Systems analysis and design in a changing world* (7th ed.). Boston, MA: Cengage Learning.

Schalles, C. (2012). *Usability evaluation of modeling languages*. New York: Springer.

Scheer, A.-W. (1998). *Business process engineering reference models for industrial enterprises*. New York: Springer, Berlin.

Scheer, A.-W., Jost, W., & Öner, G. (2007). A reference model for industrial enterprises. In P. Fettke & P. Loos (Eds.), *Reference Modeling for Business Systems Analysis* (pp. 167–181). Hershey, PA: Idea Group Publishing.

Schekkerman, J. (2004). *How to survive in the jungle of enterprise architecture frameworks: Creating or choosing an enterprise architecture framework*. Trafford, Victoria.

Schütte, R. (1998). *Grundsätze ordnungsmässiger Referenzmodellierung: Konstruktion konfigurations- und anpassungsorientierter Modelle*. Wiesbaden: Gabler.

Shields, J. L. (2006). Organization and cultural change. In W. B. Rouse (Ed.), *Enterprise Transformation: Understanding and Enabling Fundamental Change* (pp. 79–106). Hoboken, N.J.: Wiley-Interscience.

Stachowiak, H. (1973). *Allgemeine modelltheorie*. Wien: Springer.

Stair, R. M., & Reynolds, G. W. (2012). *Fundamentals of information systems*. Boston: Course Technology/Cengage Learning.

Stamatelatos, M., Grida Ben Yahia, I., Peloso, P., Fuentes, B., Tsagkaris, K., & Kaloxylos, A. (2013). Information model for managing autonomic functions in future networks. In D. Pesch, A. Timm-Giel, R. A. Calvo, B.-L. Wenning, K. Pentikousis (Eds.), *Mobile Networks and Management* (pp. 259–272). Springer International Publishing, Cham.

The Open Group. (2011). TOGAF Version 9.1. Zaltbommel: Van Haren Publishing.

Thiry, M. (2010). *Program management, fundamentals of project management*. Gower, Farnham, Surrey, England: Burlington, VT.

Thomas, O. (2005). Das Modellverständnis in der Wirtschaftsinformatik: Historie, Literaturanalyse und Begriffsexplikation, Veröffentlichungen des Instituts für Wirtschaftsinformatik im Deutschen Forschungszentrum für Künstliche Intelligenz. Inst. für Wirtschaftsinformatik im Dt. Forschungszentrum für Künstliche Intelligenz (DFKI).

Thomas, O. (2006a). Understanding the term reference model in information systems research: History, literature analysis and explanation. In C. Bussler & A. Haller (Eds.), *Business Process Management Workshops* (pp. 484–496)., Lecture Notes in Computer Science Berlin, Heidelberg: Springer.

Thomas, O. (2006b). *Management von Referenzmodellen: Entwurf und Realisierung eines Informationssystems zur Entwicklung und Anwendung von Referenzmodellen*. Berlin: Logos.

Thomas, O. (2007). Reference model management. In P. Fettke & P. Loos (Eds.), *Reference Modeling for Business Systems Analysis* (pp. 288–309). Hershey, PA: Idea Group Publishing.

TM Forum. (2011). Information framwork (SID): Information framework primer (GB922 0-P), Version 9.5. ed.

TM Forum. (2012a). Business services (Contracts) concepts and principles (GB942CP), Version 3.2. ed.

TM Forum. (2012b). Business process framwork (eTOM): Working together: ITIL and eTOM (GB921 Addendum W), Version 11.5.0. ed.

TM Forum. (2015a). Business process framwork (eTOM): Concepts and principles (GB921 CP), Version 15.0.0. ed.

TM Forum. (2015b). Information framwork (SID): Concepts and principles (GB922), Version 15.0.0. ed.

TM Forum. (2015c). Frameworks release 15.0.0: Release notes (RN354), Version 15.0.0. ed.

TM Forum. (2015d). Application framwork (TAM): Concepts and principles (GB929 CP), Version 14.5.1. ed.

TM Forum. (2015e). Business process framwork (eTOM): End-to-end business flows (GB921 Addendum E), Version 15.0.0. ed.

TM Forum. (2015f). Application framwork: The digital services systems landscape (GB929 Addendum D), Version 14.5.1. ed.

Urbaczewski, L., & Mrdalj, S. (2007). A comparison of enterprise architecture frameworks. *Issues in Information Systems, 7*, 18–23.

Van Den Berg, M., & Van Steenbergen, M. (2006). *Building an enterprise architecture practice*. Netherlands, Dordrecht: Springer.

van't Wout, J., Waage, M., Hartman, H., Stahlecker, M., & Hofman, A. (2010). *The integrated architecture framework explained*. Springer Berlin Heidelberg, Berlin, Heidelberg.

vom Brocke, J. (2003). *Referenzmodellierung: Gestaltung und Verteilung von Konstruktionsprozessen*. Berlin: Logos-Verl.

vom Brocke, J. (2007). Design principles for reference modeling: Reusing information models by means of aggregation, specialisation, instantiation, and analogy. In P. Fettke & P. Loos (Eds.), *Reference modeling for business systems analysis* (pp. 47–75). Hershey, PA: Idea Group Publishing.

Winter, R., & Fischer, R. (2007). Essential layers, artifacts, and dependencies of enterprise architecture. *Journal of Enterprise Architecture, 2*, 7–18.

Winter, R., & Sinz, E. J. (2007). Enterprise architecture. *Information Systems and e-Business Management, 5*, 357–358. doi:10.1007/s10257-007-0054-0

Winter, R., vom Brocke, J., Fettke, P., Loos, P., Junginger, S., Moser, C., et al. (2009). Patterns in business and information systems engineering. *Business & Information Systems Engineering, 1*, 468–474.

Yari, A. R., & Fesharaki, S. H. H. (2007). A framework for an integrated network management system based on enhanced telecom operation map (eTOM). In S. Ata & C. S. Hong (Eds.), *Managing Next Generation Networks and Services* (pp. 587–590). Berlin Heidelberg, Berlin, Heidelberg: Springer.

Young, L. W., & Johnston, R. B. (2003). The role of the Internet in business-to-business network transformations: A novel case and theoretical analysis. *Information Systems and e-Business Management, 1*, 73–91. doi:10.1007/BF02683511

Zachman, J. A. (1987). A framework for information systems architecture. *IBM Systems Journal, 26*, 276–292. doi:10.1147/sj.263.0276

Zachman, J. A. (1997). Enterprise architecture: The issue of the century. *Database Programming and Design, 10*, 44–53.

Zimmermann, H. (1980). OSI reference model-The ISO model of architecture for open systems interconnection. *IEEE Transactions on Communications, 28*, 425–432. doi:10.1109/TCOM.1980.1094702

Chapter 4
Designing the Architecture Solution

Abstract Designing the architecture solution combines the methodical principles in an architectural construct that offers clear recommendations for the specific challenges facing today's telecommunications operators. The result is a concrete reference solution that is presented in this chapter. First, the relevant elements are identified and arranged in an architecture structure for organization, processes, data, and applications. As an additional structural element, five industry-specific architecture domains are proposed. These architecture domains provide an overall structure of telecommunications operators. The customer-centric domain covers all architecture elements related to direct customer interactions. All technical specifics are encapsulated in the technology domain. The product domain includes the planning, development, and roll-out of new products. Both the product and the technology domain prepare the prerequisites to fulfill customer requests in the customer-centric domain. Further support activities are included in the customer domain and enterprise support domain. For each of these domains, concrete reference solutions for organization, processes, data, and applications are described and illustrated. These reference solutions combine the industry-specific TM Forum reference models and provide a detailed blueprint for the transformational needs of telecommunications operators. The reference architecture includes a hierarchical decomposition and interrelations between the different elements. Hence, this chapter presents a high-level summary of the reference architecture (cf. Sect. 4.1), an explanation of its structure (cf. Sect. 4.2), and detailed descriptions of the proposed solutions for each domain (cf. Sects. 4.3–4.6).

Today's telecommunications operators are confronted with various market and technological changes (cf. Chap. 1) that require enterprise-wide adaptations. Those adaptations are related to different parts of the enterprise. Their planning, design, and implementation are typically organized in projects and programs. Based on the concrete scenario, the as-is situation should be documented or the target picture

© Springer International Publishing AG 2017
C. Czarnecki and C. Dietze, *Reference Architecture
for the Telecommunications Industry*, Progress in IS,
DOI 10.1007/978-3-319-46757-3_4

should be designed. A modeling of certain parts of the enterprise is required. Generally these parts cover at least one of the following topics: selected business processes, mapping to the organizational structure, application functions, and/or input as well as output data. Their implementation might be related to organizational or process changes, selection of software products, and roll-out or realization of software. In most cases, those design and implementation activities are performed in different parallel projects (e.g., optimization of a CRM system, reorganization of a call center, realization of a convergent production system). A major challenge is an aligned view of all parts and their interrelations in an overall architecture.

From a methodical perspective, information systems modeling (cf. Sect. 3.1), enterprise architectures (cf. Sect. 3.2), and enterprise transformation (cf. Sect. 3.5) provide general guidance. From the content perspective, reference models (cf. Sect. 3.4) are used as recommendation for industry-specific or functional parts. In the telecommunications industry, the reference models eTOM, SID, and TAM provide industry-specific recommendations for processes, data, and applications. In practice, the identification of relevant parts of the enterprise as well as the mapping and customization of the relevant reference models are required. For this purpose, a specific reference architecture for telecommunications operators is provided in this chapter. The proposed reference architecture is an industry-specific reference model for an enterprise architecture of a telecommunications operator. It combines different existing reference models and provides a recommendation for their usage. Reference architectures are a well-accepted instrument to provide recommendations for a specific industry, such as the Y-CIM model for the production industry (Scheer 1997, p. 93; Scheer et al. 2007, p. 172).

Figure 4.1 illustrates how the different topics of the reference architecture are organized. A high-level overview of the different layers, domains, and reference process flows is provided in Sect. 4.1. A detailing of the reference architecture follows afterwards. First, the detailed structure of the process layer, organizational mapping, data layer, and application layer is explained in Sect. 4.2. According to this structure, the content is provided based on the defined architecture domains: customer-centric domain (cf. Sect. 4.3), technology domain (cf. Sect. 4.4), product domain (cf. Sect. 4.5), and customer domain as well as support domain (cf. Sect. 4.6).

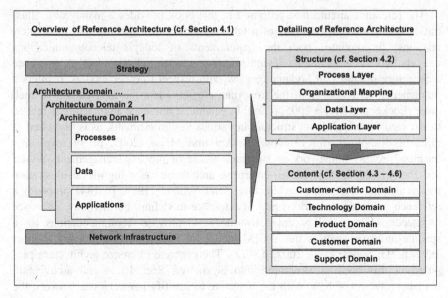

Fig. 4.1 Designing the architecture—overview

4.1 Overview of Reference Architecture

A reference architecture for telecommunications operators should provide a manageable view of all relevant parts of the enterprise in order to support the planning, design, and implementation of the architectural transformation. The requirements for this transformation are derived from the specific challenges of today's telecommunications industry (cf. Chap. 2).

Recommendations with different topical focus are provided by reference models. TM Forum Frameworx is focused on the telecommunications industry and offers industry-specific recommendations for processes (cf. Sect. 3.4.1), applications (cf. Sect. 3.4.3), and data (cf. Sect. 3.4.2) (TM Forum 2015a). ITIL is independent of any concrete industry and provides best practices for the IT service management (cf. Sect. 3.4.4) (Orand 2013). Those different reference models can be seen as pieces of a puzzle that must be selected and arranged for a specific implementation. An own empirical analysis of 184 real-life projects in the telecommunications industry shows that, in practice, TM Forum Frameworx is often used in a disintegrated manner (Czarnecki et al. 2012). Yet, it is precisely the overall understanding of the inter-relations between the architectural elements of a telecommunications operator which is essential for addressing today's transformational challenges. Telecommunications companies endeavor to integrate their organizations, applications, and technologies throughout their various mobile and fixed networks, telephony and broadband products, as well as transmission and content services.

The reference architecture proposed in this book provides a manageable illustration of all relevant elements of a telecommunications operator and their interrelations. It combines both the requirements of today's telecommunications markets and the content of relevant reference models (cf. Fig. 4.2). A reference architecture provides an organizing view with respect to other models. It offers a fundamental terminology of the important elements and the interrelation to each other (Becker and Meise 2005, p. 99). In general, a reference architecture is related to the need for a high-level structure in various design elements, as is discussed in different disciplines: for example, Becker und Meise (2005, p. 99) explain a business process framework as structural frame of process management, Winter and Fischer (2007, p. 8) see an enterprise architecture as a high-level illustration that is detailed in further architectures, and Schütte (1998, p. 184) proposes a reference model framework as part of reference modeling. Furthermore, reference architectures are a well-accepted instrument to provide recommendations for a specific industry, such as the Y-CIM model for the production industry (Scheer 1997, p. 93; Scheer et al. 2007, p. 172). The concrete reference architecture proposed in this book is structured into layers (cf. Sect. 4.1.1) and architecture domains (cf. Sect. 4.1.2). With the objective to identify interrelations between the different architectural elements, the definition of reference process flows plays an important role (cf. Sect. 4.1.3).

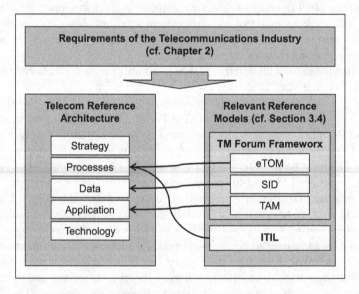

Fig. 4.2 Conceptual basis of the reference architecture

4.1.1 Layers of Reference Architecture

The reference architecture (cf. Fig. 4.3) is structured into the following five layers, which are based on the common approaches of Enterprise Architecture Frameworks (cf. Sect. 3.2.2): (1) strategy, (2) processes, (3) data, (4) applications, and (5) network infrastructure.

Strategy Layer

The *strategy layer* is influenced by the general challenges of the telecommunications industry. In the end, all transformational efforts are judged by their contribution to the overall strategic targets (Böhmann et al. 2007, pp. 128–131). In the past, business models in the telecommunications industry were based on long-term investments for network infrastructure, which was financed by stable and usage-based customer relations (Mikkonen et al. 2008, p. 180). The privatization of formerly government-owned telecommunication providers combined with new network technologies and licenses have caused tremendous competitive pressure in the telecommunications industry (Brock 2002, p. 45; Church and Gandal 2005, pp. 119–120; Economides 2005, p. 375 ff; Wellenius and Townsend 2005, pp. 575–576). In addition, the usage-based tariffs have given way to flat-rate tariffs, and the former technical view on network capacities is replaced by content-focused usage. Both require a new understanding of revenue streams combined with the need for innovative products. As a result, today's

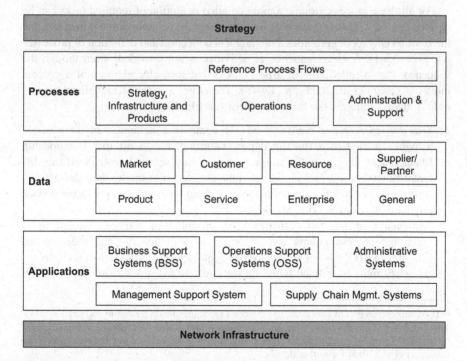

Fig. 4.3 Layers of the reference architecture

telecommunications operators are confronted with creating new strategies to address very opposing targets: efficiency increase and product innovations that require a comprehensive approach combining stagnating markets with new high-tech markets. Possible solutions could involve a complete reengineering of product development processes and systems, thus allowing faster time-to-market of innovative products and costs savings at the same time (Bruce et al. 2008, p. 15). This example shows that the strategies related to efficiency and innovations are highly dependent on the underlying processes and applications. For this reason, the clear definition of those strategic targets is an important prerequisite of the architectural transformation.

Process Layer

The *process layer* is structured according to the reference model eTOM, which provides a specific recommendation for the telecommunications industry. On a high level, processes are differentiated into the following three process groups (TM Forum 2015a, p. 9):

1. *Strategy, Infrastructure and Products (SIP)* contains all processes required to plan, design, and implement a business strategy, technical infrastructure and products.
2. *Operations* covers all processes to run a telecommunications operator with existing infrastructure and products.
3. *Enterprise Management* covers all supporting processes which are not directly involved in the core value creation.

For all these process groups, a more detailed definition of required tasks can be derived from eTOM (cf. Table 4.1). In this context, it is important to remember that the basic eTOM model provides a hierarchical collection and definition of processes (cf. Sect. 3.4.1). A clear sequence of activities is not provided[1] even though the sequence—or so-called control aspect—is an indispensable element of a process model (Axenath et al. 2005, pp. 47–48). Therefore, a concrete implementation of eTOM always requires the following steps (cf. Fig. 4.4):

1. *Selection of relevant activities*: eTOM contains a definition of all possible activities related to a functional process group, e.g., Customer Relationship Management. In a concrete usage scenario, only selected processes from different functional process groups are relevant. As an example, the sales activity from Customer Relationship Management should work together with the service provisioning from Service Management.
2. *Sequencing of selected activities*: The grouping of activities according to functional process groups is comparable to a dictionary. The order of the activities in eTOM is not related to their sequence. This sequence of the selected processes must be defined based on the concrete usage scenario.

---

[1]eTOM Addendum E (TM Forum 2010) provides a recommendation for end-to-end process flows. These end-to-end process flows were developed as an extension to the hierarchical eTOM model by the eTOM working group. The two authors of this book were involved in this development. See Czarnecki et al. (2013) for further details.

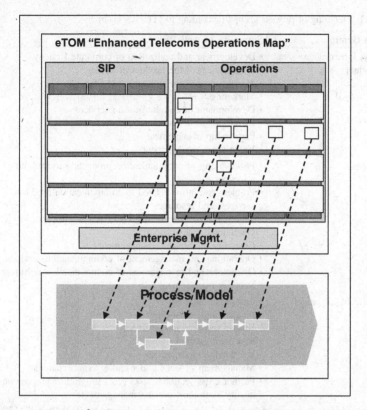

Fig. 4.4 Usage of eTOM[2]

The process layer of the reference architecture contains end-to-end process flows as an overarching view providing the sequence of eTOM activities. The interrelations between the different enterprise elements are especially important for the success of an architectural transformation. These interrelations are provided by the end-to-end process flows. They were already described in Czarnecki et al. (2013). In this book, the end-to-end process flows are further described and used as a structural element for the reference architecture.

Data Layer
The *data layer* defines the overall structure of the major data entities of a telecommunications operator. Those data entities are linked to the input and output of the processes. An aligned data model is an important prerequisite for the overall process and application model. The TM Forum reference data model SID (cf. Sect. 3.4.2) provides the definition of major data entities as well as detailed diagrams for those entities. On the highest level, SID distinguishes between eight

---

[2]Own illustration that was already published in TM Forum (2015e, p. 10).

Table 4.1 Detailing of process groups (according to ITU 2007a, b)

Process Group	Summary of Activities
Strategy, Infrastructure and Product	• Development and realization of corporate strategy • Management of the product lifecycle • Management of the infrastructure lifecycle • Marketing • Development of communication services • Development of communication resources • Partnership management • Value chain management
Operations	• Provisioning of communication products, service, and resources • Problem resolution related to communication products, services, and resources • Billing and revenue management • Operational sales support • Customer relationship management • Operations and management of communication services • Operations and management of communication resources • Supplier management
Enterprise Management	• Enterprise planning • Finance and controlling • Human resource management • Risk management • Knowledge management • Management of stakeholder and external relations • Further cross-sectional tasks (e.g., program management, quality assurance, process management)

The table provides a summary of the high-level process groups defined by eTOM (e.g. ITU 2007a, b)

domains which are used as reference for the definition of the data layer (TM Forum 2015b, p. 9):

1. *Market* contains data that is necessary to understand and address the market.
2. *Customer* encompasses all data that is relevant for dealing with the customer, such as defining, contacting, serving, and billing of customers.
3. *Product* covers all data that is related to the product lifecycle and range from definition of products to product usage.
4. *Service* contains data that is necessary for the whole service lifecycle and range from service definition to service operations.
5. *Resource* covers all data that is required for the resource lifecycle from resource definition to resource operations.
6. *Supplier/Partner* includes all data that is required for interactions with suppliers and partners from planning data to operational transactions.
7. *Enterprise* covers further specific data relevant for the enterprise.
8. *General* contains data that is required to define general data objects used in the above data entities.

In most cases, historically grown applications and processes are strongly influenced by specific characteristics and restrictions of their initial usage purpose. Consequently, specific criteria are directly stored in their structures and applications; for example, a system for selling a fixed line telephone access contains the relevant parameters for this specific product in a hardcoded manner. Accordingly, changes in existing processes and structures often require changes in program code. As an example, the launch of a new product might require up to 18 months until it is implemented in all systems (Bruce et al. 2008, p. 15). With respect to the highly competitive telecommunications markets and the increasing demand of innovative products (cf. Chap. 1), more flexible processes and applications are required. In this context, the decoupling between specific criteria and hardcoded implementations is a must. Processes and applications are developed as independently as possible with regard to the concrete customers, products, and network technologies. All specifications are then stored in the data structure.

The development of a flexible data structure for the definition and provisioning of telecommunication products is based on common approaches of the manufacturing industry (Bruce et al. 2008, p. 15). A product is specified by a bill of material listing all components and their quantity. The production plan combines this component-based view with the required production activities and their logical order. Product changes do not always require changes to processes and systems. They can easily be realized by changing data entries of a concrete bill of material or production plan.

This concept has now been transferred to the telecommunications industry. Telecommunication products do have specifics which make their definition quite complex. They are a mixture of tangible and intangible components. Some of them are dependent on location, technologies, and current capacities. The solution is a split into a tripartite structure consisting of product, service, and resource (Bruce et al. 2008, p. 19). A product is offered to a market and sold to customers. It contains commercial and technical specifications. A product consists of one or more services. Services provide functionalities, which are defined independently of their technical realization—e.g., voice telephony is a service that can be realized on various fixed or mobile network components. The technical realization of a service is specified by its resources. This tripartite structure provides a reduction of complexity by a decoupling between market-oriented products and technology-restricted resources. Furthermore it avoids mixing commercial and technical views. The customer buys a product with defined functional parameters. He is normally not interested in its technical realization. Buying a concrete product is processed in a customer order, which is then disaggregated in one or more work orders. This interrelation between those basic entities is illustrated in Fig. 4.5. It is an essential prerequisite for the definition of flexible processes and applications.

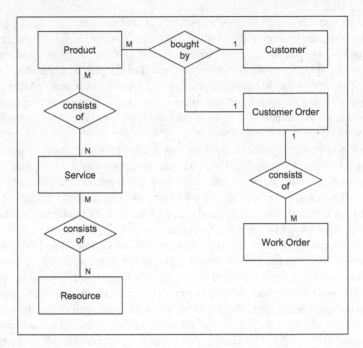

Fig. 4.5 Flexible data structure for product definition and provisioning (according to TM Forum 2015b, pp. 42–46)[3]

Application Layer

The *application layer* defines the functional structure for software systems. It is linked to the requirements defined in the strategy and process layers. In this context, the reference model TAM (cf. Sect. 3.4.3) offers a hierarchical structure of functions that are grouped according to similarities of invocation context, end user perspective, and application purpose. On the other hand, these criteria are not sufficient to identify typical application systems or system groups used in telecommunications operators.

In practice and in science, the differentiation between *Operations Support Systems* (OSS) and *Business Support Systems* (BSS) is widely accepted (Bruce et al. 2008, p. 15; Choi and Hong 2007, p. 3012; ITU 2008b, p. 8; Kelly 2003, p. 109; Mikkonen et al. 2008, p. 181; Snoeck and Michiels 2002, p. 331). Both can be seen as groups of systems. OSS are those systems that support the operations and maintenance of telecommunication networks (ITU 2008b, p. 8). In the past, those systems were designed as proprietary and monolithic software products offered by network suppliers (Lewis 2001, p. 242; Misra 2004, p. 3). They were directly linked to the network elements and technologies (Bruce et al. 2008, pp. 17–18). Later, the

---

[3]Own illustration summarizes the relations between selected data entities based on SID (TM Forum 2015b, pp. 42–46). The cardinalities are based on the notation proposed by Chen (1976).

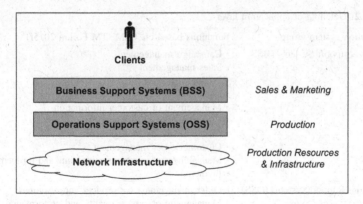

Fig. 4.6 Differentiation between BSS and OSS

integration of network elements—e.g., from different network suppliers—was required. Therefore, today's OSS also include software products that operate and manage network elements from different suppliers and technologies.

BSS cover systems that are necessary to manage and offer products, as well as to answer customer requests and to react on reported problems. These systems include the typical sales and marketing functions. A Customer Relationship Management (CRM) system is a typical software system that is part of the BSS.

In summary (cf. Fig. 4.6), OSS can be seen as production systems that manage the production resources and infrastructure. BSS provide the link between those production systems to the clients. The interrelation between both BSS and OSS is one of the major success factors for an efficient mastering of the technological and market-related complexity. In an ideal architecture, OSS encapsulate all technical details of the production. On the other hand, the BSS encapsulate all details related to the clients. The reference architecture proposed in this book offers a detailed concept for a functional breakdown of BSS and OSS as well as their interfaces.

In addition to BSS and OSS, further application systems are required to support the corporate management, the supply chain, as well as the administrative and support activities. Those application systems are grouped into *Management Support Systems*, *Supply Chain Management Systems*, and *Administrative Systems*. In Table 4.2, the different functions of the application systems groups are defined based on the reference model TAM.

Network Layer

The *network layer* contains the physical components that are required to operate the relevant telecommunication networks. In the past, a major challenge for telecommunications operators was the technical realization of telecommunication networks. The building-up and operating of a telecommunications network are an enormous infrastructure investment. The physical network infrastructure is still an important factor for telecommunications operators and markets (Sharkey 2002, pp. 180–204). In most countries, it is not only seen as a production factor of a single company but also as

Table 4.2 Detailing of application layer

Application system group	Functions based on TAM (TM Forum 2015f)[a]
Business Support Systems (BSS)	• Campaign management • Sales management • Product management • Management of customer self-service • Management of customer information • Customer contact management • Customer loyalty management • Customer order management • Management of customer complaints and problems • Invoicing
Operations Support Systems (OSS)	• Order management for services and resources • Management of service quality and performance • Management of service problems and resource problems • Management of service catalogs and service inventories • Management of resource inventories • Management of resource lifecycle • Collection and rating of usage and billing data
Administrative Systems	• Finance and controlling • Human resource management • Asset management • Revenue assurance • Knowledge management • Security management • Management of regulatory and legal requirements • Further administrative functions (e.g., travel management)
Management Support Systems	• Collection and monitoring of performance indicators • Analysis and evaluation • Reporting • Planning and decision support
Supply Chain Management Systems	• Procurement management • Supplier management • Partner management • Settlement of interconnect and roaming

[a]The definition of functions is based on the high-level application groups of TAM (TM Forum 2015f). The table summarizes these functions and provides a mapping to the application system groups

an important country-wide economic factor which requires governmental regulation. For a long time, telecommunications operators were organized around their network infrastructure. The systems for operating the network were offered as specific solutions by the network suppliers (Lewis 2001, p. 242; Misra 2004, p. 3). Processes and other systems had to fit to those technology-specific network operations systems. The entire business was developed in a bottom-up manner starting with the network infrastructure. This has led to silo-oriented processes and systems (Bruce et al. 2008, p. 16). Such structures are inflexible and not suitable for the changed market conditions, as time-to-market for new products of more than 18 months shows (Bruce et al.

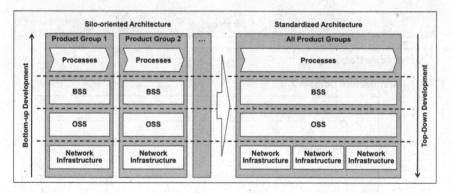

Fig. 4.7 Interrelation between network infrastructure and standardized architecture (according to Czarnecki 2013, p. 149; Czarnecki and Spiliopoulou 2012, p. 395)

2008, p. 15). The required flexibility is realized by a decoupling of network infrastructure, processes and applications (Bertin and Crespi 2009, pp. 187–190; Bruce et al. 2008, p. 16; Knightson et al. 2005, p. 49).

The reference architecture proposed in this book supports this concept (cf. Fig. 4.7). Technology-neutral processes are defined from an end-to-end perspective. The applications are standardized based on functional groupings. From a technical perspective, a separation between transport and communication services is required in order to provide standardized interfaces between the network infrastructure and the OSS. This separation is supported by a Next Generation Network (NGN) (Choi and Hong 2007, p. 3005; Knightson et al. 2005, pp. 50–52; Yahia et al. 2006, p. 16). However, in most cases, different technical specifics on the network infrastructure layer are still necessary (e.g., different network suppliers). These technical specifics are stored in flexible data structures.

4.1.2 Defining Architecture Domains

The major elements of the reference architecture for telecommunications operators can be structured into these layers: processes, data, and applications (cf. Sect. 4.1.1). There are interrelations between the layers; for example, the sales processes are related to the product data model that is implemented in the Business Support Systems (BSS) and aligned with the Operations Support Systems (OSS). Furthermore, the standardization of processes requires the encapsulation of certain technical details; for example, changes of the network infrastructure should be realized through changes of data entries rather than changes of processes and applications. Therefore, it is recommended that the interrelations between the different layers already be considered at a high level during the architecture design. However, the complexity of telecommunications operators requires a clear structure in order to efficiently manage the interrelations between the different architecture

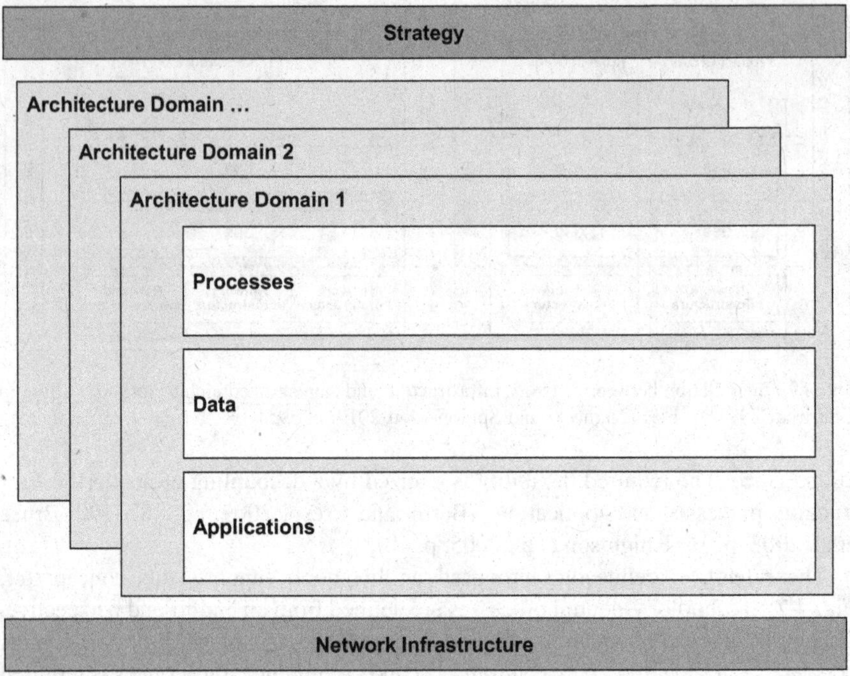

Fig. 4.8 Structuring according to architecture domains

elements. In this book, a topical structure based on the value chain of telecommunications operators is proposed. This high-level structure is called architecture domains. Each architecture domain contains processes, data, and applications that are relevant for this domain (cf. Fig. 4.8).

The architecture domains are based on a topical structuring of telecommunications operators. The core activities of a telecommunications operator can be summarized as follows:

- Sales and provisioning of telecommunication products,
- Customer service regarding complaints and technical problems.

These core activities require the availability of telecommunication products as well as a network infrastructure, i.e.:

- Development and launch of telecommunication products,
- Development and realization of telecommunication services,
- Roll-out, extension, operations, and maintenance of network infrastructure.

In addition, further support activities are required:

- Supporting marketing activities—e.g., market research, marketing campaigns,
- General and administrative activities—e.g., finance and human resource management.

In a first step, the aforementioned activities can be structured into (1) those activities that are directly involved in the value creation and (2) those activities supporting the value creation (Becker et al. 2005; Porter 2004). The so-called primary activities are directly involved in the value creation with respect to products and external customers. Those primary activities are supported by so-called support or secondary activities (Becker et al. 2005; Porter 2004).

For a telecommunications operator, the primary activities are sales, provisioning, and customer service related to telecommunication products. As a prerequisite, a network infrastructure and telecommunication services are needed. These are technology-related support activities. The development and launch of products according to market requirements are product-related support activities. With respect to the high competition in telecommunications markets, additional marketing-related activities are responsible for understanding and pushing market demands. All three groups of support activities are specific to the telecommunications industry. In addition, there are general support activities (e.g., finance) that are required for almost every company, irrespective of a specific industry. This structure of the activities of a telecommunications operator is illustrated in Fig. 4.9.

This structure forms the basis for the definition of architecture domains for telecommunications operators. The architecture domains should provide a general structure independent of the concrete implementations of processes and systems. In the discipline of architecture development, Aier and Winter (2008, pp. 180–182) propose the usage of domains. In the telecommunications industry, Snoeck and Michiels (2002, pp. 333–334) have already used the domains *employee*, *product*, *order*, and *configuration* for structuring the development of BSS and OSS. Also, the TM Forum uses domains for their reference models based on the data entities (cf. Sect. 3.4).

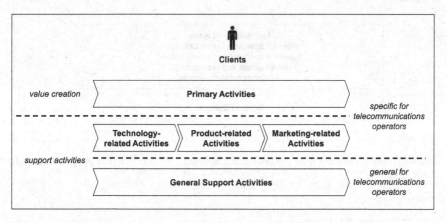

Fig. 4.9 Structuring the activities of a telecommunications operator

According to the aforementioned structuring of activities, the following domains are proposed (Czarnecki 2013; Czarnecki et al. 2013; Czarnecki and Spiliopoulou 2012; TM Forum 2015d, 2010)[4]:

- *Customer-centric domain* contains all primary activities, such as sales and customer service. These processes are defined from an end-to-end perspective always starting and ending with the customer.
- *Technology domain* covers the roll-out, extension, operations, and maintenance of the network infrastructure as well as the development and realization of telecommunication services.
- *Product domain* contains the development and launch of telecommunication products based on the services provided by the technology domain.
- *Customer domain* focuses on marketing activities, such as market research or campaigns. In contrast to the customer-centric domain, the processes of the customer domain support customer-related activities, such as preparing successful sales through marketing campaigns.
- *Support domain* contains all general support activities, such as finance or human resource management.

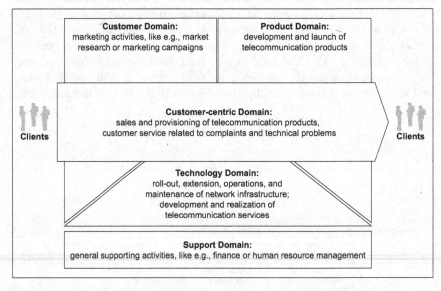

Fig. 4.10 Architecture domains (according to Czarnecki 2013, p. 109; TM Forum 2010, p. 8, 2015e, p. 12)

---

[4]The proposed domains are based on intensive project work, research and development. This work was mainly conducted by Detecon in various different teams. The two authors had leading roles in this development. The domains were already published in prior scientific publications (e.g. Czarnecki 2013; Czarnecki et al. 2013) and white papers (e.g. TM Forum 2015d, 2010).

These domains are illustrated in Fig. 4.10. Please see Czarnecki et al. (2013) for more details on their development process. They are the outcome of various projects in the telecommunications industry. They were contributed to the TM Forum eTOM working group. After confirmation by the TM Forum, they were officially included in the eTOM standards in GB921-E (TM Forum 2015d)[5] (cf. Sect. 3.4.1).

4.1.3 Defining Reference Process Flows

The architecture domains provide a high-level structure of a telecommunications operator. The next step is a detailing of those domains through reference process flows. The aim of those reference process flows is the overall illustration of the relevant activities for each domain. This is the basis for understanding the inter-relations between the different domains. In addition, it provides first indications of the required data elements and applications.

The definition of reference process flows should still be on an aggregated level and from an end-to-end perspective. The end-to-end perspective means that for a specific use case, the complete sequence of activities is described: for example, in the customer-centric domain an end-to-end process covers the selling of a telecommunication product from the customer contact to the billing.

The TM Forum reference model eTOM (e.g. TM Forum 2015a) is a well-accepted industry standard for processes in the telecommunications industry (cf. Sect. 3.4.1). Compliance to eTOM is an indispensable requirement for the process definition. The original eTOM standard provides a hierarchical collection of processes (Kelly 2003). The end-to-end processes require a sequencing of the activities provided by eTOM—i.e., the control aspect (Axenath et al. 2005, pp. 47–48) is added to the process hierarchy provided by eTOM. As a solution, so-called reference process flows were developed (Czarnecki et al. 2013). The development of those reference process flows was an iterative and joint effort of the eTOM working group based on the results of various real-life projects. Please see Czarnecki et al. (2013) for a detailed description of this development process. Today, the reference process flows are a well-accepted part of the eTOM reference model and published in GB921-E. In the following, there is a description according to Czarnecki et al. (2013), Czarnecki (2013, pp. 106–134), and GB921-E (TM Forum 2010, 2015d)[6] of these reference process flows as part of the reference architecture.

---

[5]The domains were first published in 2010 in the document GB921-E as part of the eTOM standard (TM Forum 2010). A revised version of GB921-E was published in 2015 (TM Forum 2015d).

[6]The reference process flows are based on intensive project work, research and development. This work was mainly conducted by Detecon in various different teams. The two authors had leading roles in this development. The reference processes flows were already published in prior scientific publications (e.g. Czarnecki 2013; Czarnecki et al. 2013; Czarnecki and Spiliopoulou 2012) and white papers (e.g. TM Forum 2010, 2015d).

The reference process flows are developed independent of a concrete organizational structure and with a strict focus on the activities of a telecommunications operator. The starting points are the use cases relevant for each of the defined process domains. In fact, on an aggregated level, all activities can be summarized in a manageable amount of use cases.

The *customer-centric domain* should cover all use cases that are invoked by a direct request or requirement from a customer. They can be summarized as follows:

- Requesting information,
- Buying a product (from the customer perspective, i.e., selling from the enterprise perspective),
- Using a product,
- Changing an existing contract,
- Terminating an existing contract,
- Reporting a technical problem,
- Reporting a commercial complaint.

The *technology domain* decouples the commercial product from its technical realization. It contains all uses cases which are related to the development, operations and maintenance of the network infrastructure. They can be summarized as follows:

- Technical realization of a product,
- Solving of a technical problem,
- Technical monitoring of product usage,
- Management of services,
- Management of resources,
- Management of network capacity,
- Management of network continuity.

The *product domain* covers the whole product management lifecycle, which is divided into the following use cases:

- Idea generation and management of the product portfolio,
- Development and launch of new products,
- Changing existing products,
- Termination of existing products.

The *customer domain* focuses on marketing activities which are, in most cases, related to customers but unlike the customer-centric domain, not invoked by them. The reference process flows are covering the following two use cases:

- Management of the customer relation,
- Management of sales activities.

The *support domain* includes general support activities. They are not specific to the telecommunications industry. According to eTOM, the following activities are identified (TM Forum 2015a):

- Managing the corporate strategy and planning,
- Managing finance and controlling,
- Managing the organization and human resources,
- Procuring of goods and services,
- Managing the supply chain,
- Managing corporate risks,
- Managing knowledge and research,
- Managing quality, processes and performance,
- Managing projects and programs,
- Managing external relations and stakeholders,
- Managing corporate communications,
- Managing legal and regulatory topics.

On an aggregated level, all the above use cases are sufficient to cover most of the activities of a telecommunications operator. They are used as input for the development of the reference process flows. A mapping between the use case and the reference process flows is shown in Table 4.3. According to the end-to-end logic, most of them are named by a starting event and closing event: for example, *Request-to-Answer* covers all activities starting from receiving a request to answering this request. However, this logic could not be applied for all reference process flows. The supporting activities include some parts which follow a functional logic, e.g., *Capacity Management*. In this case, the name does not include start and end events, but the process is still defined from an activity viewpoint for a whole use case. The process *Capacity Management*, for example, includes all capacity-related activities from definition of capacity targets to operational realization of additional capacities.

According to the typical categorization of business processes (Becker et al. 2005; Porter 2004), the domains are differentiated in activities which are directly related to the value creation and those which are supporting the value creation (cf. Figs. 4.10 and 4.11). Due to the complexity of telecommunications markets and technologies, the second structural element is a decoupling between the commercial and technical view (Bertin and Crespi 2009, pp. 187–190; Bruce et al. 2008, p. 16; Knightson et al. 2005, p. 49). The commercial view deals with customers and products. From a technical view, customers are seen as subscribers and products are realized by services and resources (Bruce et al. 2008, p. 16). Resources are related directly to the technical infrastructure. Services define a specific functionality realized by one or more resource(s) and offered as (part of) one or several product(s) (Bruce et al. 2008, p. 19; Czarnecki and Spiliopoulou 2012, p. 393; ITU 2008a; TM Forum 2015b, p. 46).

Table 4.3 Mapping between use cases and reference process flows

Process domain	Use case	Reference process flow[a]
Customer-centric domain	Requesting information	Request-to-Answer
	Buying a product	Order-to-Payment
	Using a product	Usage-to-Payment
	Changing an existing contract	Request-to-Change
	Terminating an existing contract	Termination-to-Confirmation
	Reporting a technical problem	Problem-to-Solution
	Reporting a commercial complaint	Complaint-to-Solution
Technology domain	Technical realization of a product	Production-Order-to-acceptance
	Solving of a technical problem	Trouble-ticket-to-solution
	Technical monitoring of product usage	Usage-to-usage-data
	Management of services	Service lifecycle management
	Management of resources	Resource lifecycle management
	Management of network capacity	Capacity management
	Management of network continuity	Continuity management
Product domain	Idea generation and management of the product portfolio	Idea-to-business-opportunity
	Development and launch of new products	Business-opportunity-to-launch
	Changing existing products	Decision-to-Relaunch
	Termination of existing products	Decision-to-elimination
Customer domain	Management of the customer relation	Customer relation management
	Management of sales activities	Sales Management
Support domain	Managing the corporate strategy and planning	Strategic and corporate management
	Managing finance and controlling	Financial management
	Managing the organization and human resources	Human resource management
	Procuring of goods and services	Supply chain management
	Managing the supply chain	
	Managing corporate risks	Enterprise risk management
	Managing knowledge and research	Enterprise effectiveness management
	Managing quality, processes and performance	
	Managing projects and programs	
	Managing external relations and stakeholders	Corporate communications
	Managing corporate communications	
	Managing legal and regulatory topics	Legal and regulatory management

[a]The development of the reference process flows was a joint effort of the eTOM working group based on various projects and Detecon's knowledge development. Both authors of this book were involved in these initiatives. The reference process flows of the customer-centric and technology domains were published in TM Forum GB921-E (2010, 2015e) and are described here accordingly. The product, customer and support domains are not published in GB921-E. Their description here is based on the results of various projects and Detecon's knowledge development. In addition, the reference process flows of the customer-centric, technology, and product domain are described on a high level in Czarnecki et al. (2013).

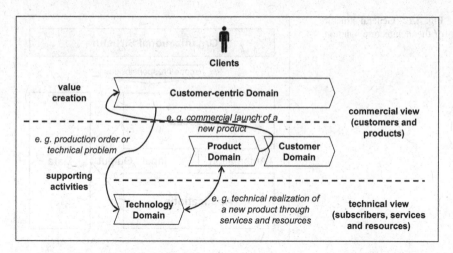

Fig. 4.11 Interrelation between reference process flows

There are various interrelations between the different domains which need to be considered in the design of a concrete architecture (cf. Fig. 4.11). The value creation is covered by the customer-centric domain. It deals with all customer-related use cases from a commercial view. All technical topics related to subscribers, services, and resources are forwarded to the technology domain. The development and launch of products are covered by the product domain. New products are handed over for operations from the product domain to the customer-centric domain. The technical realization of products through services and resources is handled by the technology domain. Marketing activities are covered by the customer domain and might result in concrete customer requests, which are then handled by the customer-centric domain. These general differentiations and inter-relations between the domains are the basis for the detailed description of the different architecture domains (cf. Sect. 4.3-4.6).

4.2 Structuring the Architecture Solution

The architecture solution is structured into different layers which define the processes, applications, and data from a conceptual viewpoint (cf. Fig. 4.12). In addition, the organizational structure is considered, as it provides a solid indication for the general scope of a concrete enterprise. For the architecture design and implementation, the mapping to the organizational structure is important in order to define responsibilities (cf. Sect. 5.1.4).

Processes are an indispensable part of an information system architecture (e.g. Hammer and Champy 1994; Scheer et al. 2007). The interrelation between processes and applications should be seen as a mutual dependency (e.g. Laudon and

Fig. 4.12 General structure
of the architecture solution

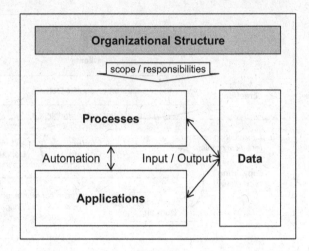

Laudon 2012). On the one hand, processes define requirements for applications; on the other hand, applications enable the improvement of processes. Both are supported by an overall data model that defines input and output for processes as well as applications.

The purpose of an architecture solution is the transformation to a target situation that creates a value for the enterprise (Böhmann et al. 2007, p. 129). Hence, the structure of this architecture solution should support its implementation. The implementation is differentiated in the organizational and technical implementation. The organizational implementation focuses on changes of the future processes and their mapping to the organizational structure. The objective of the technical implementation is the automation of processes through adequate software systems. In today's telecommunications markets, high flexibility and fast implementations are combined with tremendous cost pressure (cf. Sect. 2.1). Therefore, the usage of standardized software products is an important objective in most cases.

With respect to most enterprise architecture frameworks (e.g. Matthes 2011; Schekkerman 2004; Urbaczewski and Mrdalj 2007), the design of the architecture solution starts with the process layer. For the organizational implementation, those processes are mapped to the organizational structure. For the technical implementation, a mapping between processes and logical application elements are provided. Those application elements are not dependent on concrete software systems. Both the processes and applications are supported by an overall logical data model.

The content for these different parts of the architecture solutions is defined based on the different TM Forum reference models that provide specific recommendations for processes, data, and applications for the telecommunications industry. Those reference models are structured in a hierarchical manner using different levels. The usage of those reference models requires a consistent structure which combines the different content and their interrelations. This structure is described according to the different layers of the reference architecture: process layer (cf. Sect. 4.2.1), data layer (cf. Sect. 4.2.3), and application layer (cf. Sect. 4.2.4). In practice, the mapping between the reference architecture and the organizational structure is essential. Therefore, this topic is explained in Sect. 4.2.2.

4.2.1   Structuring the Process Layer

Process modeling is typically organized in a hierarchical manner (Rosemann 2003). On a high level, a process framework provides an overview of all processes that are relevant for a company (Becker and Meise 2005, p. 123). A well-known high-level framework is Porter's value chain (Porter 2004). It provides the major activities that are required to create the company's outcome. Those activities are then further detailed in a strict hierarchical manner (cf. Fig. 4.13). As an example, the outbound logistics can be further detailed in different process steps—e.g., goods receipt and unpacking. Those process steps can be then further detailed: for example, goods receipt is further detailed into receive goods, inspection of incoming goods, and goods receipt posting.

A process framework should be mutually exclusive and collectively exhaustive (Betz 2011, p. 73)—i.e., each activity covers all required aspects and there are no overlaps between activities. In the above example, the activity *inbound logistics* includes all process steps that are required, and it is clearly separated from the activity *operations*. The detailing of a process step always provides a clear decomposition on a more detailed level. However, on a more detailed level, single process steps might be repeated in different activities: for example, the process step *billing* might be included in both activities *marketing and sales* as well as *customer service*. This repetition is recommended as it supports the process standardization. There should be no difference in processing bills in *marketing and sales* compared to *customer service*.

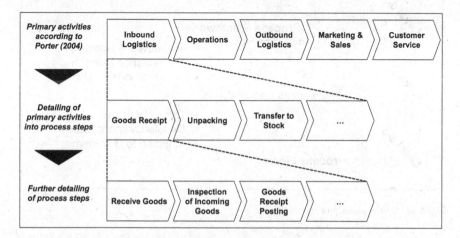

Fig. 4.13 General concept of hierarchical process structure

Even though the hierarchical detailing of processes is well accepted (e.g. Rosemann 2003), a common definition and naming of the different process hierarchies does not exist. Therefore, as a starting point, this book uses the definition of process levels proposed by the reference process model eTOM (cf. Sect. 3.4.1). In eTOM, a detailing of processes from level 0 to level 5 is defined (TM Forum 2015a, p. 42). However, the process detailing does not encompass any process flow relation (TM Forum 2015f, p. 11). The relation between different process elements is an indispensable requirement of every process design (Axenath et al. 2005). In this book, a process structure that combines the eTOM levels with practical implementation requirements is proposed as follows (cf. Fig. 4.14):

- *Level 0–1* are linked to the process framework. They provide an overall view of the enterprise combined with first process groupings (TM Forum, 2015a, p. 42). The recommended process framework consists of the architecture domains (level 0) and reference process flows (level 1) (cf. Sect. 4.3.1). In addition, the process framework should be mapped to the organization to provide a sponsorship on the top-management level.
- *Level 2* provides a first detailing of the reference process flows (level 1). It includes process description, objectives, activities, performance indicators, and involved organizational entities.
- *Level 3–5* covers process flows on an operational level. On level 3 it starts with the sequencing of activities according to the process descriptions (level 2). Each activity can then be further detailed on level 4 and level 5. However, a detailing of all processes on level 4 or level 5 is not mandatory. The required level of detail depends on the specific situation. In addition, the process flows contain a detailed mapping to the organizational responsibilities and IT systems.

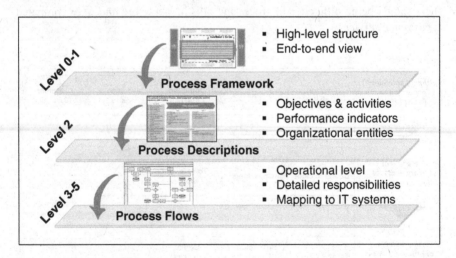

Fig. 4.14 Definition of process levels

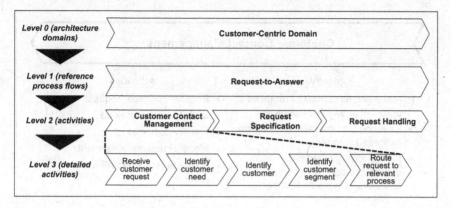

Fig. 4.15 Example of process levels

The following is a concrete example according to the above structure (cf. Fig. 4.15). On a high level, processes for telecommunications operators are structured according to the architecture domains (level 0) and reference process flows (level 1). Both are described in detail in Sect. 4.3.1. In this example, the *customer-centric domain* was chosen. It groups all processes that are directly initiated by the customer. The reference process flow *request-to-answer* is one process out of the customer-centric domain. It describes the handling of requests. In a first step, the *request-to-answer* process is divided into the following activities (level 2): *customer contact management, request selection and specification,* and *request handling.* Those level 2 activities are further detailed on level 3. As example, *customer contact management* is divided into *receive customer request, identify customer need, identify customer, identify customer segment,* and *route request to relevant process.*

For the level 2 process, the objectives, further detailing of activities, organizational mapping, and performance indicators are described (cf. Fig. 4.16). In this example, *customer contact management* is further detailed in a level 2 process description. This description provides a manageable overview of the process. The objectives are used as an indication for the planning and prioritization of processes. The organizational mapping is an important input for the implementation and execution. The performance indicators are used for the monitoring of the process execution and as an input for continuous improvement. In addition, the list of activities provides further details about the process.

The level 2 process description does not provide an exact logical flow. In this example, it is not defined if the activity *identify customer* is always performed or only for selected requests. These details are provided by the level 3 process flows (cf. Fig. 4.17). The process flows should be designed in a process flow notation, e.g., Business Process Model and Notation (BPMN) (OMG 2011), Event-Driven Process Chains (EPC) (Scheer 2000), or subject-oriented Business Process Management (S-BPM) (Fleischmann 2013). On the other hand, there are further

> ## Customer Contact Management

Objectives	Activities
• Managing all contacts / requests of potential or existing customers • Ensuring that interactions conform to agreed standards • Standardizing customer experience across all contact channels, regions and products	• Receive customer request • Identify customer need • Identify customer • Identify customer segement • Re-route request to relevant process
Involved Organizational Entities	**Performance Indicators**
• Customer • Customer Care • Sales	• Number of contacts by channel / Total number of contacts • Average customer wait time to be served • Average customer request closure time • No. of contacts that are closed on first contact / Total no. of contacts

Fig. 4.16 Example of level 2 process description

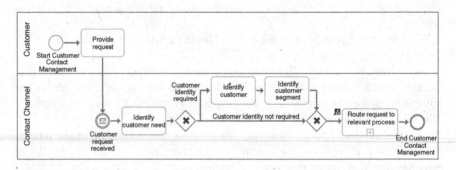

Fig. 4.17 Example of level 3 process flow

notations that could be considered in a practical context. Please see, e.g., Long (2014, pp. 6–13) for a comparison of different process modeling notations. The selection of a process modeling notation and a supporting tool is based on the specific requirements (Davies and Reeves 2010). In this book, BPMN 2.0 (OMG 2011) is used with swim lanes defining the organizational responsibilities.

In the example, the level 3 process flow is defined for the level 2 process *customer contact management*. In this process, the customer and the respective contact channel are involved. The identification of the customer and the customer segment is not performed for all requests. Based on the request type, the request is routed to the relevant process. This is defined as a sub-process, which is further detailed on level 4.

4.2.2 Structuring the Organizational Mapping

A company's organization is a tool that coordinates people's action to achieve a certain outcome (Jones 2013). The organizational development as well as organizational theories about interrelations with strategy and technology are major business administration disciplines (e.g. Cummings and Worley 2009; McAuley et al. 2007; Tsukas and Knudsen 2005). In this book, only a short summary of the relevant theoretical basics is provided.

Organizational content can be differentiated as follows (Jones 2013; Laudon and Laudon 2012):

- Organizing and coordinating the activities that are required to create a company's outcome. This part is defined by the *processes*.
- Organizing people by a formal system of authorities and responsibilities. This part is defined by the *organizational structure*.

Processes describe the *How*, while the organizational structure defines the *Who*. It is common sense that processes are an indispensable part of information systems (e.g. Becker 2011; Hammer and Champy 1994; Porter 2004). There is a strong relation between processes and applications, as both transform a certain input into an output (Laudon and Laudon 2012). In most enterprise architectures, the major business part is formed by processes (Aier et al. 2011; Schekkerman 2004; The Open Group 2011) supported through reference process models (e.g., eTOM). On the contrary, the organizational structure varies in most practical implementations (Mansfield 2013, p. 67). A major aspect of organizational structures is the division of labor, which can be influenced by the number of hierarchical levels, functional specialization, differentiation, departmentalization, and divisionalization (Mansfield 2013, p. 67). Even though the discipline of organizational design provides tools to rationalize decisions about the organizational structure (Jones 2013) such as the ideal number of hierarchical levels, in practice, a concrete organizational structure is heavily influenced by the power structure and politics of the involved individuals (Laudon and Laudon 2012).

For the telecommunications industry, the reference process model eTOM provides a consensus for business processes that is confirmed by major industry players (cf. Sect. 3.4.1). A comparable reference for the organizational structure does not exist. The consideration of the organizational structure is indispensable for every architectural transformation. The organizational structure defines the responsibilities of the transformation itself as well as the execution of the future target processes. In a practical context, a clear consideration of the organizational structure might decide about success or failure.

In process management, the idea of a process organization is discussed (e.g. Becker 2011; Hammer and Champy 1994). A process organization describes the shift from a functional organization to a process-oriented organization (Becker 2011)—i.e., labor is not assigned through functional specialization, but rather through activities required to create a certain output. As a result, a complete reorganization is proposed as part of the process implementation (e.g. Becker 2011; Hammer and Champy 1994). In practice, most telecommunications operators have undergone various reorganizations. Proposing a reorganization as a mandatory part of an architectural transformation creates an unnecessary complexity which might even lead to a failure of the whole transformation. Therefore, the new architecture is only linked to the organizational structure through a mapping. In a first step, the architecture is developed independently of a concrete organizational structure. Through the mapping, those responsibilities which are important for a successful implementation are then defined. Any organizational changes would not impact the architecture itself, but only the organizational mapping. This concept is comparable with a matrix structure (e.g. Jones 2013).

A concrete organizational mapping can be structured according to the following dimensions (cf. Fig. 4.18):

- *Phase* provides a mapping to the phases of the architectural transformation. This might vary based on the chosen enterprise architecture framework and enterprise architecture management (cf. Sect. 3.1). Based on common approaches (Aier et al. 2011; The Open Group 2011), those phases could be differentiated into planning, development, implementation, execution, and monitoring.
- *Role* defines the role of the organizational entity in the architectural transformation. Those roles vary based on the chosen project management methodology (e.g. Dinsmore and Cabanis-Brewin 2014; Westland 2007). In this book, the following roles are proposed: steering, communication, decision, approval, expert input, realization, and execution.
- *Hierarchical level* provides a mapping to the organizational hierarchy. Differentiations can be made between top management, middle management, and operational level (Laudon and Laudon 2012).
- *Cross-functional level* distinguishes between end-to-end and functional responsibilities. Assuming a matrix structure (e.g. Jones 2013), the end-to-end responsibility assures the alignment between the different functional entities which are reflected by the functional responsibilities.

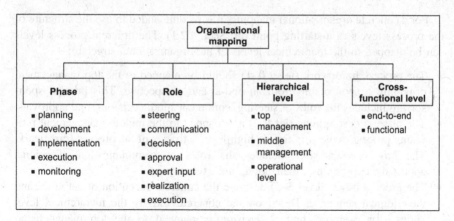

Fig. 4.18 Dimensions of organizational mapping

Those characteristics of the different dimensions are influenced by each other. The steering, for example, is typically conducted by the top management while execution is the responsibility of the operational level. Even though the organizational mapping is one of the most important success factors of an architectural transformation, a concrete mapping requires the consideration of the specific situation. Based on the experience with various architectural transformations the following points are recommended:

- Planning, development, and implementation are typically conducted in a project organization which performs a hand-over to the day-to-day business for execution and monitoring. It is essential to already involve the responsible personnel from the day-to-day business in the project organization.
- An involvement of the top management is important as an architectural transformation affects the whole enterprise. Still the detailed development, implementation, and execution is managed by the middle management and conducted on an operational level. The right balance between the hierarchical levels is essential, especially for steering, communication, decisions, and approvals.
- Typically approvals are required between the different phases of the architectural transformation—e.g., the development of a certain process is approved before its implementation. It is essential to clearly define the exact proceeding and the involved personnel at the beginning of the architectural transformation. The number of personnel should be manageable, and they should be involved in the development as well as the implementation.
- The value of an architectural transformation is linked to the cross-functional alignment between processes, applications, and data. Nevertheless, in most cases, the organizational responsibilities are defined in a functional structure. Therefore, both the end-to-end and functional view are essential for a successful implementation.

For a concrete organizational mapping, it is recommended to use the structure of the process levels as a starting point (cf. Sect. 4.2.1). The different process levels can be mapped to the hierarchical levels of the organizational structure:

- The process framework (level 0–1) should be mapped to the top management providing a sponsorship from an end-to-end perspective. This process sponsorship includes the roles of steering, communication, decisions, and approvals.
- The process descriptions (level 2) are mapped to the middle management providing process ownership for the high-level activities of the process framework. The process ownership includes the roles of communication, decisions, approvals, expert input, realization, and execution.
- The process flows (level 3–5) describe the concrete execution of activities and their logical sequence. Based on the concrete activity the hierarchical level varies—for example, budget decision is mapped to the top management, answering a customer phone call is mapped to the operational level. In addition, the process flows include a mapping to the relevant applications and defines the responsibilities for the realization from the application perspective.

According to those hierarchical levels, the organizational mapping is structured as follows (cf. Fig. 4.19). Based on the process framework, sponsors from the top management are defined, e.g., by architecture domain. Process owners from the middle management are nominated from an end-to-end perspective to coordinate the cross-functional alignment of high-level processes (level 2). They are supported by sub-process partners, and these are defined based on the functional responsibilities of the detailed process activities. All of these responsibilities are related to the process content. Furthermore, a responsibility for the methodical governance

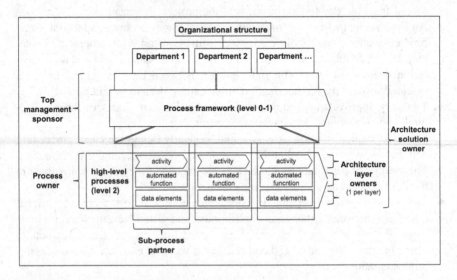

Fig. 4.19 Proposal for structuring the organizational mapping

should be defined. A dedicated architecture solution owner is responsible for the overall governance of the architecture. Typically, this is a special organizational entity, such as an architecture management department. In addition, architecture layer owners are responsible for an overall alignment of the different architecture layers (e.g., processes, applications).

In a concrete transformation project, a nomination of persons for the different parts of the organizational mapping is required. The related tasks and required effort will vary according to the specific roles. Figure 4.20 shows an illustrative definition of tasks and effort based on a real-life transformation project. Such a definition sets the exceptions right from the beginning and helps to identify the right persons.

	Tasks	Effort
Top management sponsor	▪ Overall steering of architectural transformation ▪ Communication within the entire organization ▪ Quick response to escalations and enforcement of needed decisions	▪ Monthly steering committee ▪ Monthly status reporting
Architecture solution owner	▪ Overall steering of architectural transformation from a methodical perspective ▪ Development, communication, and implementation of methodical tools and guidelines ▪ Structuring the overall architectural transformation ▪ Definition and control of overall transformation targets ▪ Identification of overall improvement potentials and set-up of initiatives	▪ Fulltime
Architecture layer owner	▪ Accountability for design, implementation and continuous improvement of respective architecture layer ▪ Alignment of solution in their respective architecture layer ▪ Identification of functional experts in the organization of the telecommunications operator responsible for the actual execution of design, implementation and continuous improvement tasks	▪ Overall coordination ▪ Time for communication and coordination within own organization

Process owner	▪ Leadership for the related process (level 2) within the functional organization
	▪ Management of development, realization, execution, and monitoring of related process (level 2)
	▪ Overall coordination of architectural transformation:
	   ▪ Coordination and communication within own organization (vertical)
	   ▪ Coordination of cross-functional consensus (horizontal)
	▪ Decision-making, approvals and communication
	▪ Expert of related process (level 2)
		▪ Overall coordination
		▪ Access on short notice for needed escalations
		▪ Time for communication and coordination within own organization

Sub-process partner	▪ Overall responsibility related to the functional organization which is involved within the sub-process (level 3-5) ▪ Operational support of process owner related to development, realization, execution, and monitoring ▪ Expert of related sub-process (level 3-4)	▪ 20% of working time available for operational activities

Fig. 4.20 Exemplary task definitions for organizational mapping

4.2.3   Structuring the Data Layer

Data elements are linked to the process and application layers. For processes, the data elements provide a consistent structure of input and output elements. For applications. The storage, handling, and analysis of data are important requirements. Moreover,in the telecommunications industry, the data structure is indispensable to overcome historically grown and silo-oriented processes and application systems. The objective is a decoupling of processes and application systems from concrete products and technologies. These specifics are stored in the data structure (Bruce et al. 2008, p. 15). Hence, the development of an overall data model is a prerequisite for flexible processes and application systems.

The reference data model SID (cf. Sect. 3.4.2) provides a definition of specific data entities for the telecommunications industry (ITU 2008a; TM Forum 2015b). SID provides a reference data model from a business perspective. The focus is on a logical data view which is independent of any physical implementation in a database or application system. Also, SID follows a hierarchical structure (TM Forum 2015b, p. 10):

- *SID Domain* (level 0) is the highest level of aggregation. It structures the data elements according to business areas. The SID domains are linked to eTOM level 0 and TAM level 0. Hence, they assure an alignment between the data, process, and application layer.
- *Aggregated Business Entity* (ABE) (level 1–2) provides an aggregation of business entities based on their topical context, e.g., the ABE *Customer* contains all business entities that are required to describe a customer. SID differentiates between level 1 and level 2 ABEs. Level 2 ABEs are a detailing of level 1 ABEs, but still on an aggregated level.
- *Business entities* are logical data entities that are described from a business perspective. They could be tangible (e.g., customer), abstract (e.g., subscriber), or transactional (e.g., customer order).
- *Attributes* are used to further detail business entities.
- *Relations* are defined between business entities.

With this hierarchical definition, SID can be structured in a high-level illustration of major data entities and a detailed logical data model (cf. Fig. 4.21). The definition of ABEs and their relations provide an aggregated understanding of specific data entities for a telecommunications operator as an important prerequisite for the development of flexible processes and application systems. Business entities and their relations are illustrated in Entity Relationship Models (ERM) following the notation of the Unified Modeling Language (UML). These details can be used for the design of an overall logical data model as basis for the concrete implementation of application systems. The usage of standard software products might not require

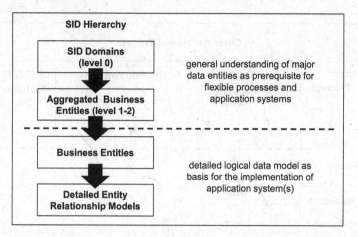

Fig. 4.21 SID hierarchy

the design of an own data model, but could utilize the predefined data model of the software product. As major software vendors in the telecommunications industry follow the TM Forum standards,[7] compliance with SID can be assumed.

For the development of an architecture solution, the high-level understanding of data entities and their relation is essential. Figure 4.22 illustrates an exemplary inter-relation between processes and data entities. The decoupling between a customer-related sales process from the network-related provisioning process is essential to manage the complexity of different network technologies (Bruce et al. 2008, p. 15). Therefore, SID defines the ABE *product* which relates to the customer-related view. Such a product consists of service(s) that are realized by resources. The ABE *resource* provides the network-related view, which is linked to the customer-related view by the ABE *service*. The required provisioning activities follow the same differentiation. They are structured into the ABE *customer order* and the ABE *work order*. This example shows that the product and technology-independent process design requires a common understanding of a high-level data model.

In order to provide an alignment between process, data, and application layers, the SID ABEs can be mapped as input and output in the process flows. The first option for this mapping is a detailed table as appendix to the process flows (cf. Fig. 4.23). The advantage of this option is that the process flows are focused on the activity sequence. Additional information can be flexibly added in this table; for example, besides the mapping to the SID ABEs, this table could also include the

---

[7]Please see the conformance certification page under www.tmforum.org which provides detailed certification results of software products.

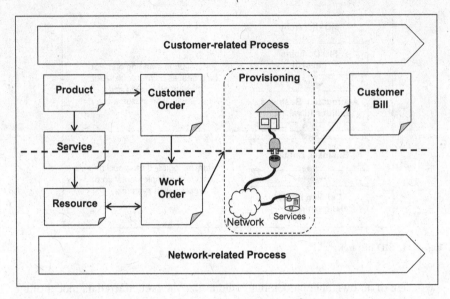

Fig. 4.22 Exemplary interrelation between ABEs and processes[8]

		real-life example
Process (level 3)	**Description**	**SID ABEs (level 1)**
Identify customer need	Customer selects preferred service option via IVR or personally at the shop. Based on the chosen service option, the sales representative identifies the customer need.	Customer Interaction
Identify customer	Sales representative identifies customer through login, password, name, unified customer ID, etc.	Customer Interaction, Customer
Identify customer segment	Customer segmentation (consumer; small / medium size business; enterprise business) is necessary in order to redirect customer to the relevant services.	Customer

Fig. 4.23 Exemplary mapping between processes (level 3) and SID ABEs

organizational responsibilities and application functions. The second option is the integration of data elements directly into the process flow diagram (cf. Fig. 4.24). This option depends on the chosen process modeling notation.

---

[8]Translated version of an own illustration published in Czarnecki (2013, p. 168).

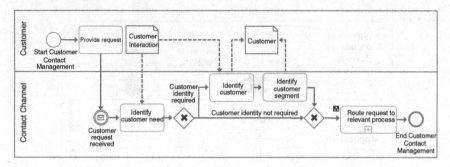

Fig. 4.24 Exemplary mapping of SID ABEs in BPMN diagram

4.2.4 Structuring the Application Layer

The application layer provides a logical view on the automation of business processes. The information system research understands information systems as a set of interrelated components consisting of people, data, software, and hardware executing processes and procedures (Laudon and Laudon 2012; Stair and Reynolds 2012, pp. 2–16). Those information systems could be either manual or supported through the use of computers and technical infrastructure (Laudon and Laudon 2012; Stair and Reynolds 2012, pp. 10–16). An application system is a concrete system automating parts of an information system through adequate software and hardware (Wigand et al. 2003, pp. 1–4).

In this book, the application layer first defines a logical view on application systems (cf. Fig. 4.25) comparable with the integration perspective described by Winter and Fischer (2007, p. 8). Processes describe the overall value creation from an organizational perspective (e.g. Becker 2011; Hammer and Champy 1994; Porter 2004). A process transforms an input into an output (Becker 2011). The required activities are defined in a hierarchical manner as well as a sequential order (cf. Sect. 4.2.1). Those activities could be supported by functions provided through application systems. Comparable to processes, functions also transform an input into an output (Laudon and Laudon 2012). Nevertheless, the grouping of processes and application functions might differ.

According to Laudon and Laudon (2012) in general the major four applications are (1) enterprise systems (collect data from different functions), (2) supply chain management systems (manage relations with suppliers), (3) customer relationship management systems (deal with customers), and (4) knowledge management systems (capture and apply knowledge). Even though a vertical and horizontal integration of application systems is a major objective (Laudon and Laudon 2012), a typical sales process might involve all the above mentioned application systems. Therefore, a logical application view is created to define and group the application functions. This logical application view is based on the TM Forum reference model TAM (cf. Sect. 3.4.3). It provides a hierarchical grouping of application functions that are specific for a telecommunications operator (TM Forum 2015c).

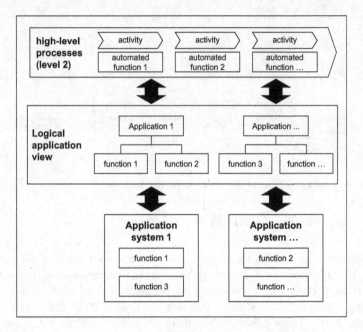

Fig. 4.25 Interrelation between processes, logical applications, and application systems

TAM groups application functions according to the criteria *invocation context,* *end user perspective,* and *purpose* (TM Forum 2015c, p. 12). The invocation context combines application functionalities that are used in the same context (e.g., dealing with a customer request). The end user perspective groups those application functions that are used by the same end user (e.g., call center agent). The purpose combines application functions that pursue a similar objective (e.g., transparency about customer relations). This grouping of application is provided in the following hierarchical structure (TM Forum 2015e, pp. 19–21):

- *TAM domains* (level 0) provide a high-level structure and are linked to the SID domains (TM Forum data reference model). An example is the *Customer Management Domain.*
- *TAM application areas* (level 1) are a first functional detailing of each domain. An example is *Customer Order Management* as an application area of the *Customer Management Domain.*
- *Further detailing* (level 2–4) provides the definition of concrete application functions. Examples are *Customer Order Establishment* and *Customer Order Orchestration* as level 2 detailing of *Customer Order Management.* On level 3, *Customer Order Establishment* is further detailed, for example, into *Channel Guidance and Data Capture* and *Customer Qualification.* For Customer Qualification on level 4 a decomposition into *Customer Credit Eligibility* and *Offering Availability* is proposed. As a consequence, level 3 and level 4 decompositions are only provided for selected application functions.

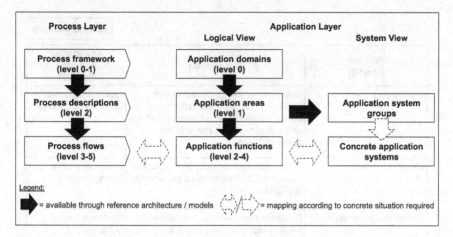

Fig. 4.26 Interrelation between process and application layers

The design of the overall architecture solution requires an alignment between the process and the application layers (cf. Fig. 4.26). The process layer provides a hierarchical structure from a process framework (level 0–1) to detailed process flows (level 3–5). Also, the logical view of the application layer is structured in a hierarchical manner from application domains (level 0) to application functions (level 2–4). For both, the hierarchical mapping is provided through the TM Forum reference models eTOM and TAM.

A concrete architecture design then requires a mapping between the detailed process flows and application functions for those activities that are automated. On the one hand, this mapping defines requirements for the implementation of concrete application systems. On the other hand, the capabilities of automated application functions could result in process improvements. In literature, this bilateral interrelation is called *duality of information technology* (Hunter 2011; Ward and Peppard 2002, p. 51; Zuboff 1988, p. 390).

For a concrete implementation, the application layer requires a link between the logical and the system view. On a generic level, the TAM application areas (level 1) are mapped to application system groups (cf. Table 4.2). The implementation in application systems is specific for a concrete situation. Assuming that a standard software product should be selected, a mapping between required application functions and concrete application systems can be utilized for this selection.

		real-life example

Process (level 3)	Description	Application Function (level 2)
Identify customer need	Customer selects preferred service option via IVR or personally at the shop. Based on the chosen service option, the sales representative identifies the customer need.	Customer Interaction Collection & Storage
Identify customer	Sales representative identifies customer through login, password, name, unified customer ID, etc.	Customer Profile Management
Identify customer segment	Customer segmentation (consumer; small / medium-sized business; enterprise business) is necessary in order to redirect customer to the relevant services.	Customer Profile Management

Fig. 4.27 Exemplary mapping between processes (level 3) and application function (level 2)

In a concrete architecture solution, the mapping between processes and application functions can be realized through a table that provides details for each process flow (cf. Fig. 4.27). In this case, the process flow focuses on the activity sequence. Further details are provided in this table, which can be easily extended for organizational mapping or data objects.

Depending on the process modeling notation used, the application function could also be included in the process flow diagram. In this book, the Business Process Model and Notation (BPMN) is used. BPMN does not provide dedicated symbols for an application mapping. However, BPMN allows the extension through additional artifacts, which could be used for a mapping of application functions as part of the process diagram (cf. Fig. 4.28).

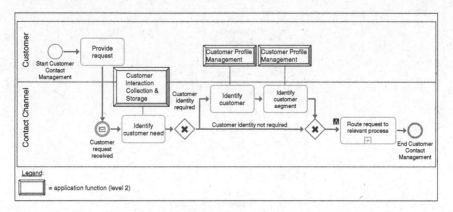

Fig. 4.28 Exemplary mapping of application functions as BPMN diagram

4.3   Detailing the Customer-Centric Domain

The customer-centric domain contains all primary activities, such as sales and customer service. These activities are defined from an end-to-end perspective, always starting and ending with the customer. The reference process flows (cf. Sect. 4.3.1) provide a recommendation for the process layer. The mapping between the customer-centric domain and the organizational structure is an essential step in a real-life transformation project (cf. Sect. 4.3.2). Furthermore, recommendations for the data layer (cf. Sect. 4.3.3) and application layer (cf. Sect. 4.3.4) are provided. The interrelations between those different layers are summarized in high-level illustrations of the customer-centric domain (cf. Sect. 4.3.5).

4.3.1   Reference Process Flows of the Customer-Centric Domain

All reference process flows[9] of the *customer-centric domain* start with an event initiated by the customer and end with an event related to the customer. Figure 4.29 provides an overview of the seven reference process flows of the customer-centric domain.

---

[9]The reference process flows are based on intensive project work, research and development. This work was mainly conducted by Detecon in various different teams. The two authors had leading roles in this development. The reference process flows of the customer-centric domain were already published in prior scientific publications (e.g. Czarnecki 2013; Czarnecki et al. 2013; Czarnecki and Spiliopoulou 2012) and white papers (e.g. TM Forum 2015e, 2010). The descriptions and illustrations in this section are a completely revised version.

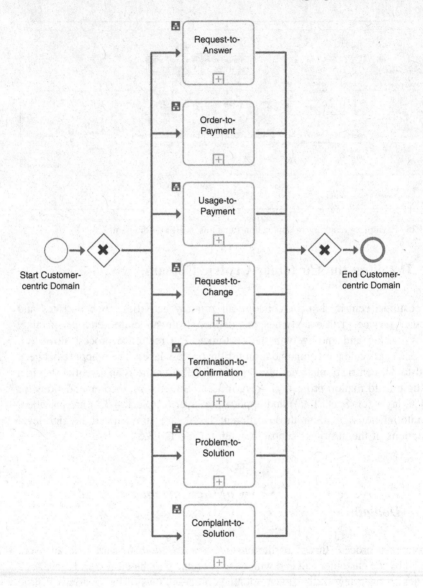

Fig. 4.29 Reference process flows of customer-centric domain (level 1)

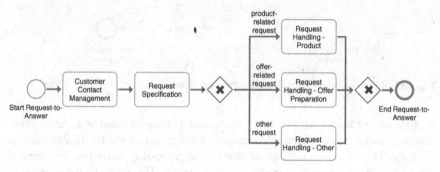

Fig. 4.30 Reference process flow Request-to-Answer (level 2)

Request-to-Answer (cf. Fig. 4.30) provides information to a customer based on his request. With respect to concrete products or contracts, this process deals with pre-sales, cross-selling, and up-selling opportunities—always, however, initiated by the customer himself. In addition, it answers general requests, such as about opening hours or location of retail shops. This process ends with an answer, which could be, e.g., a binding offer.

Order-to-Payment (cf. Fig. 4.31) is the typical sales process. It starts with the customer order. The decision of the customer to buy a concrete product was taken. This decision is either related to the Request-to-Answer process or the customer domain. The Order-to-Payment process deals with the commercial processing of a customer order, the provisioning of the product, and the billing. Depending on the product, the provisioning might include technical tasks, which are then forwarded to the technology domain.

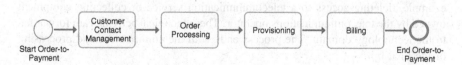

Fig. 4.31 Reference process flow Order-to-Payment (level 2)

Using a telecommunication product is a process itself, which is covered by *Usage-to-Payment* (cf. Fig. 4.32). This process starts with the customer decision to use a product based on an existing contract. It ends with the payment of the product usage. Based on the contract, the usage either requires a usage-based payment, which is normally volume or time related; or, it is part of a flat-rate agreement. A mixture of both is also common—e.g., a flat rate with a volume limit. The collection and rating of usage data is part of the technology domain. The payment itself can be pre-payment or post-payment. In addition, for complex products, the Usage-to-Payment process might include further payment-relevant services—e.g., for video-on-demand as part of a broadband TV service.

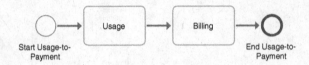

Fig. 4.32 Reference process flow Usage-to-Payment (level 2)

Request-to-Change (cf. Fig. 4.33) starts with a change request of a client. Those changes can be differentiated into changes of the customer master data or existing contracts. Depending on the type of change, the processing might require technical tasks which are forwarded to the technology domain. This could be the case for an address change of a location-based product—a fixed-line phone connection, for example. In contrast, the change of the bank account in a post-paid contract is a purely commercial activity.

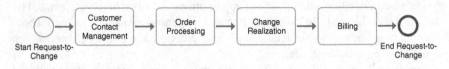

Fig. 4.33 Reference process flow Request-to-Change (level 2)

Termination-to-Confirmation (cf. Fig. 4.34) covers the termination of existing products from a commercial perspective. It starts with the termination request by the client. Depending on the corporate strategy, this process might include customer retention activities which could result in a cancellation of the termination request. In most cases, the termination requires technical activities, which could be, for example, deleting access to a telecommunication service or collecting equipment owned by the telecommunications operator. These technical activities are forwarded to the technology domain. The process ends with the confirmation and processing of the final bill.

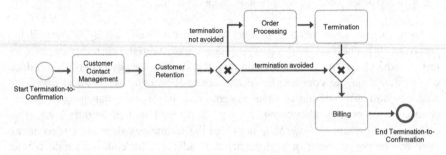

Fig. 4.34 Reference process flow Termination-to-Confirmation (level 2)

Problem-to-Solution (cf. Fig. 4.35) deals with technical problems reported by the customer. This process starts with the problem report. For technical problems, different support levels are distinguished. The customer-centric domain covers the high-level support based on well-defined scripts or tools. More sophisticated technical activities are forwarded to the technology domain. However, the overall responsibility in terms of a problem ticket ownership remains in the Problem-to-Solution process. In addition, billing activities might be involved, which could be either a credit note as compensation or the invoicing of the problem resolution. The latter could be the case if the problem was caused by customer-owned equipment not covered by the contract with the telecommunications operator.

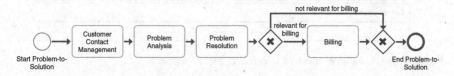

Fig. 4.35   Reference process flow Problem-to-Solution (level 2)

Complaint-to-Solution (cf. Fig. 4.36) deals with commercial complaints—i.e., all non-technical complaints. This process does not have any interface to the technology domain. The processing of the complaint depends on the complaint type and the corporate strategy. Complaints could be differentiated into complaints related to a clear legal obligation (e.g., wrong invoicing) and complaints related to customer dissatisfaction (e.g., unfriendly behavior of sales staff). Both cases might involve billing activities resulting in credit notes, either as legal obligation or as goodwill compensation.

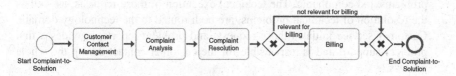

Fig. 4.36   Reference process flow Complaint-to-Solution (level 2)

4.3.2   Organizational Mapping of the Customer-Centric Domain

The organizational mapping of the seven reference process flows of the customer-centric domain depends on the concrete organizational structure, which will vary from enterprise to enterprise. In order to support a concrete organizational mapping, a description of general functional roles which can be used as a starting point for an organizational structure is given below. These general functional roles are derived from the experience with various implementation projects. They follow a functional organizational structure which is orthogonal to the process perspective.

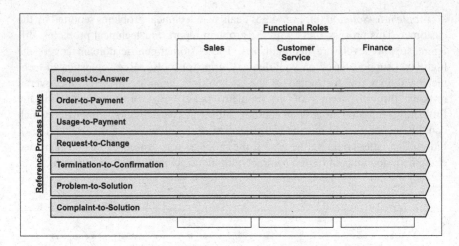

Fig. 4.37 Functional roles involved in customer-centric domain

Based on the activities of the reference process flows of the customer-centric domain, the following general functional roles are assumed (cf. Fig. 4.37):

- *Sales* is responsible for planning, managing and executing selling activities. With respect to the customer-centric domain, sales is responsible for the operational handling of customer requests and orders. The technical provisioning is part of the technology domain. The sales role initiates the provisioning and is responsible for the customer contact during provisioning.
- *Customer service* covers all after-sales activities. With respect to the customer-centric domain these are changes and termination requests as well as problems and complaints. The technical execution of those requests, as well as the resolution of technical problems, are both routed to the technology domain. Customer service initiates those requests and problem reports, provides a first problem support, and is responsible for the customer contact during the whole process.
- *Finance* is responsible for all financial activities, which can be differentiated into strategic and operational activities. With respect to the customer-centric domain, finance is responsible for operational billing activities as well as dealing with billing complaints. In both cases. There is an interface to sales or customer service, which manages the customer contact during the whole process.

The definition of organizational entities for these roles depends on various criteria, which can be structured as follows (cf. Fig. 4.38):

- *Channel* defines the sales and contact channels used by the enterprise. Typical channels are call center, shop, Internet, sales representatives, and indirect sales.
- *Customer types* describe the types of customers addressed by the enterprise. They can be differentiated into consumer, small and medium enterprise, corporate, and wholesale.

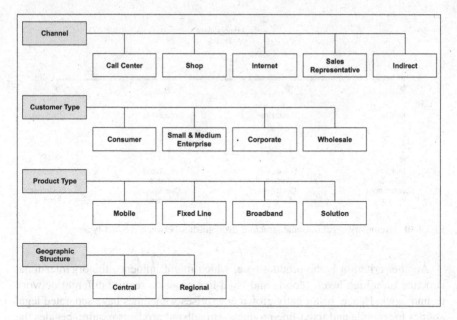

Fig. 4.38 Criteria of organizational scope in the customer-centric domain

- *Product types* define the types of products offered by the enterprise. A typical categorization is mobile, fixed line, broadband, and complex solutions.
- *Geographic structure* is related to the geographic distribution of organizational responsibilities. It can be divided into central and regional structures. A regional structure depends on the concrete geographic conditions and is defined by factors such as the number of regions.

According to the characteristics of the above criteria, the relevant functional roles of sales, customer service, and finance are designed in different ways. In the following, a description of exemplary specifications of organizational designs is given.

The skills required for serving consumer customers is, in most cases, completely different from the skills required for corporate customers and/or wholesale customers. Therefore, organizational designs might distinguish between consumer, corporate, and wholesale customers on a high level (cf. Fig. 4.39).

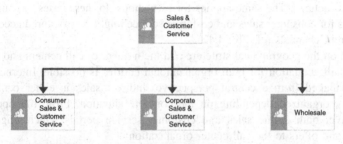

Fig. 4.39 Exemplary organizational structure distinguishes between customer types

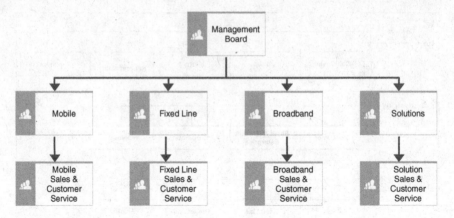

Fig. 4.40 Exemplary organizational structure distinguishes between product types

Another criterion is the product type, which might influence the organizational structure on a high level. Mobile and fixed-line products require different network technologies. Hence, historically grown enterprises sometimes have separated legal entities for mobile and fixed-line products. Broadband products require, besides the network infrastructure. Additional services (e.g., Internet access) which were originally offered by Internet Service Providers (ISPs). Solutions for the communication requirements of corporate customers are a complex business which includes planning, consulting, developing, implementing, and operating of solution-based products. Therefore, a possible organizational structure could distinguish between the different product types on a high level (cf. Fig. 4.40). In this case, each product-based division contains its own sales and customer service units.

A possible mixture between the distinction of the customer type and the product type is shown in Fig. 4.41. In this case, the sales and customer service for consumer, small and medium enterprise, and wholesale is structured according to these different customer types. Those different units cover the responsibility for sales and customer service of all product types offered to these customer types. In this example, solution-based products are offered to corporate customers. They form a separate organizational unit.

Major contact channels for consumer customers are call centers, shops, Internet, and indirect sales. If own call centers are operated, they require complex organizational structures. The same applies for own shops. In these cases, organizational structures for consumer sales and customer service might be structured according to the different channels (cf. Fig. 4.42).

Based on the geographical structure and the number of call centers and shops, a regional differentiation for both organizational entities is possible. Internet is normally managed from a central perspective. Indirect sales is either centrally or regionally organized depending on the concrete situation. From the operational perspective, both define sales and customer service portals which might route requests and offers to the call center organization.

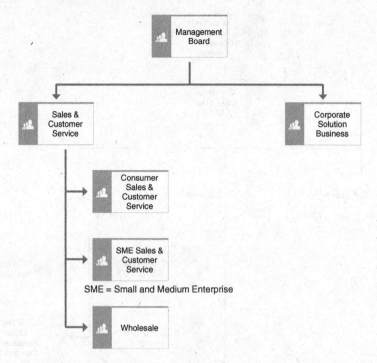

Fig. 4.41 Exemplary organizational structure combines distinction between customer and product type

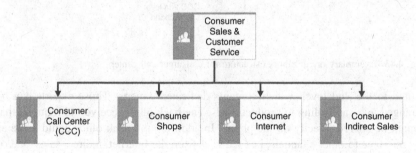

Fig. 4.42 Exemplary organizational structure for consumer sales and customer service based on channels

For the consumer part of the customer-centric domain, the call center organization plays a decisive role. A possible organizational design of the consumer call center is to structure it into 1st level, 2nd level, and back office (cf. Fig. 4.43). The 1st level answers all calls. These calls could be related to all product types and requests. Hence, the 1st level focuses on answering typical requests based on clearly defined scripts. If the request cannot be solved by the 1st level, it is then

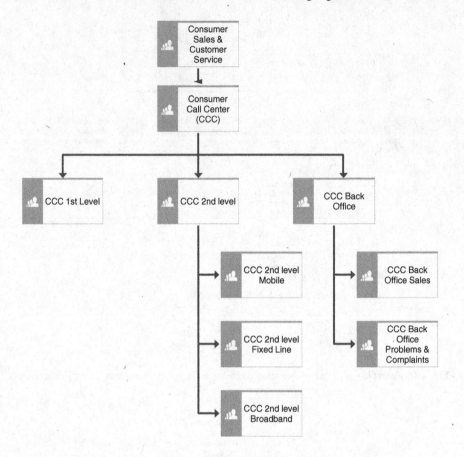

Fig. 4.43 Exemplary organizational structure for consumer call center

routed to the 2nd level, which contains dedicated experts. Those experts can be arranged, e.g., according to product types. The back office receives all requests that cannot be solved directly on the phone. In addition, the back office could serve as the interface to other channels, i.e., requests from the Internet, indirect sales, and/or shops might be routed to the back office. A possible structure of the back office is to distinguish between sales requests, problems, and complaints.

In contrast, the sales and customer service of corporate customers is normally organized based on a key account management. In this case, sales representatives are the major contact channel. They could be organized by industries, regions, or technologies and supported by a central sales support. The after-sales service could be realized by a helpdesk which offers a technical problem support. Figure 4.44 shows an exemplary organizational structure for corporate sales and customer service, which is based on a regional structure for the sales representatives.

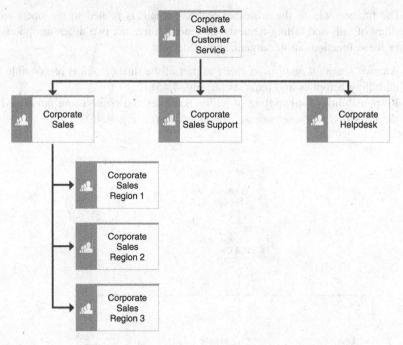

Fig. 4.44 Exemplary organizational structure for corporate sales and customer service

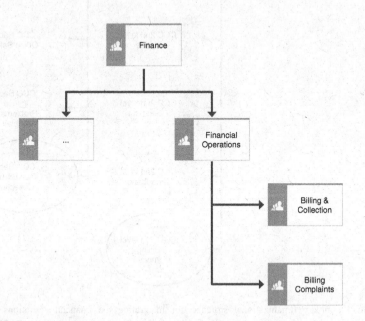

Fig. 4.45 Exemplary organizational structure of own financial operations entity

The finance role of the customer-centric domain is related to the operational handling of bills and billing requests. In general, there are two different options to cover these functions in the organizational design:

1. An *own financial operations entity* as part of the finance unit is responsible for all billing activities and requests (cf. Fig. 4.45).
2. Responsibilities for handling of billing activities and requests are *integrated in the sales and customer service organizations* (cf. Fig. 4.46).

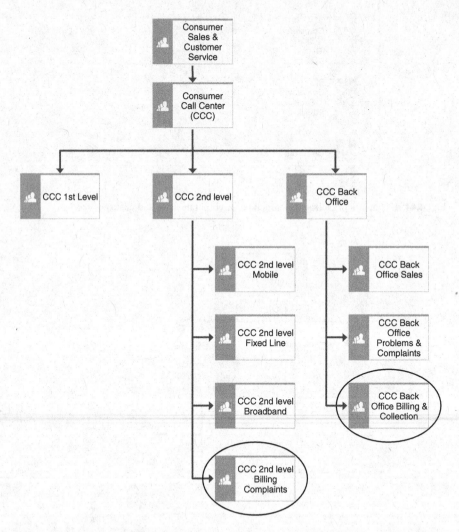

Fig. 4.46 Exemplary organizational structure for integrating the financial operations in the consumer call center

For both options, different concrete realizations are possible. In the case of integrating the financial operations into the sales and customer service organization, the specific design might vary based on the above explained criteria; for example, it could be realized through a central entity or a regional structure. A further example would be a differentiation of billing activities by customer types. An additional criterion could be the billing type which varies between pre-paid and post-paid. Also, a mixture between integrating and own entity is possible. A common realization would be the integration of billing complaints in the call center organization and the central handling of billing and collection in the finance unit.

4.3.3   Data Layer of the Customer-Centric Domain

The data processed in the customer-centric domain is mainly related to the TM Forum SID domains *customer* and *product* (TM Forum 2015b, pp. 19–24). The separation between the customer view and the production view is also realized in the data layer. Products are composed of services that are realized by resources (Bruce et al. 2008, p. 19; Snoeck and Michiels 2002, p. 335). In this context, it is essential to clearly understand the term *service* as defined by SID. In the meaning of SID, services are any part of a product which could be tangible (e.g., a mobile phone) or intangible (e.g., a broadband connection).[10] Products are bought by customers through customer orders, which are realized in the technology domain through work orders (cf. Fig. 4.47).

The reference model SID contains detailed entity relationship models for all SID domains (e.g. TM Forum 2015b, pp. 37–48). Besides those detailed models, a high-level understanding of the major data entities is an important basis for the architecture design. Therefore, in the following there is a summary of the major Aggregated Business Entities (ABEs) that are relevant for the customer-centric domain.

The SID domain *product* includes all required data structures for the definition, offering, pricing, and usage of products. The major relevant ABEs are (TM Forum 2015b, pp. 19–22):

- *Product Specification* (ABE level 1) defines the general functions and characteristics of products offered to customers. It is a blueprint of a concrete product sold to a customer. A product specification is created in the product process domain. In the customer-centric domain, the product specification is used as input to offer concrete products.
- *Product Offering* (ABE level 1) is based on a product specification which is offered to a certain market via a certain sales channel at a certain price.
- *Product* (ABE level 1) is a concrete instance of a product offering that is sold to a concrete customer. The product is related to a certain realization, which might contain details about configuration and location.
- *Product Usage* (ABE level 1) contains usage statistics related to a product. Those statistics are used for billing purposes.

---

[10]Please see Sect. 3.3.4 for differentiation between the term *service* in SID and ITIL.

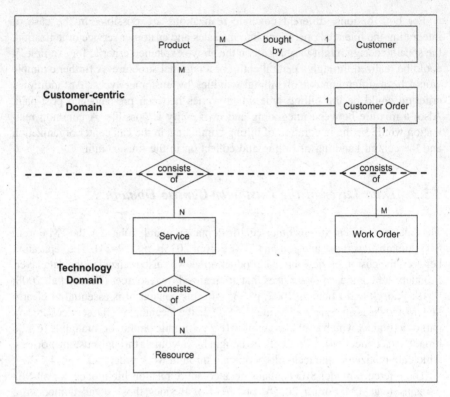

Fig. 4.47 Differentiation of SID domains between customer-centric and technology domain (according to TM Forum 2015b, pp. 42–46)[11]

The SID domain *customer* includes all required data structures for managing customer interactions, orders, problems, and bills. The major ABEs are (TM Forum 2015b, pp. 22–24):

- *Customer* (ABE level 1) contains static data about the customer, such as contact details, account details, preferences, and credit profile.
- *Customer Interaction* (ABE level 1) stores all relevant interactions with the customer across all contact channels—for example, an interaction with a call center agent to report a technical problem. The interaction can require the creation or change of further ABEs, such as Customer Problem (ABE level 1).
- *Customer Order* (ABE level 1) is a concrete order of a product offering by a customer through a customer interaction. A Customer Order is created and handled in the customer-centric domain and realized through work orders in the technology domain.

---

[11]Own illustration summarizes the relation between selected data entities based on SID (TM Forum 2015b, pp. 42–46). The cardinalities are based on the notation proposed by Chen (1976).

- *Customer Problem* (ABE level 1) contains a concrete problem reported by a customer through a customer interaction. A customer problem is reported in the customer-centric domain. If it cannot be solved there, it is then forwarded to the technology domain.
- *Customer Bill* (ABE level 1) contains the relevant data for billing products of customer. The customer bill is linked to further ABEs that are relevant for billing, such as Applied Customer Billing Rates (ABE level 1).

4.3.4 Application Layer of the Customer-Centric Domain

The processes of the customer-centric domain are linked to the Business Support Systems (BSS) (Bruce et al. 2008, p. 15; Kelly 2003, p. 109; Snoeck and Michiels 2002, p. 331). BSS cover the typical sales and marketing functions. They support all processes that are related to selling products and serving customers. A Customer Relationship Management (CRM) system is a typical software system that is part of the BSS. The technical realization of those products is part of the technology domain, which is linked to the Operations Support Systems (OSS) and network infrastructure (Bruce et al. 2008, pp. 17–18). For the customer-centric domain, the technical production through OSS and network infrastructure can be seen as a black box that offers its services through (a) standardized interface(s) (cf. Fig. 4.48).

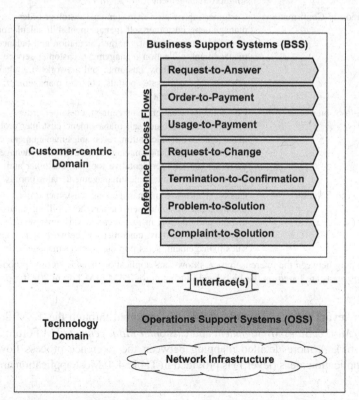

Fig. 4.48 Mapping of customer-centric domain to BSS

Table 4.4 Mapping between reference process flows and application areas (level 1) for customer-centric domain

Reference process flows	Application areas (Level 1) (TM Forum 2015f, pp. 22, 63)[a]
Request-to-Answer	Customer information management, customer order management, customer self management, customer service representative toolbox, customer and network care, channel sales management, sales portals, contract management, solution management
Order-to-Payment	Customer information management, customer order management, customer self management, bill calculation, receivables management, charge calculation and balance management, collection management, customer service representative toolbox, customer and network care, channel sales management, sales portals, contract management, solution management
Usage-to-Payment	Bill calculation, receivables management, charge calculation and balance management, collection management
Request-to-Change	Customer information management, customer order management, customer self management, bill calculation, receivables management, charge calculation and balance management, collection management, customer service representative toolbox, customer and network care, channel sales management, sales portals, contract management, solution management
Termination-to-Confirmation	Customer information management, customer order management, customer self management, bill calculation, receivables management, charge calculation and balance management, collection management, customer service representative toolbox, customer and network care, channel sales management, sales portals, contract management, solution management
Problem-to-Solution	Customer information management, customer order management, customer self management, customer problem management, bill calculation, receivables management, charge calculation and balance management, collection management, customer service representative toolbox, customer and network care, channel sales management, sales portals
Complaint-to-Solution	Customer information management, customer order management, customer self management, billing inquiry, dispute and adjustment management, customer service representative toolbox, customer and network care, channel sales management, sales portals, contract management

[a]The mapping between the reference process flows and application areas is an own proposal. The application areas are based on the high-level illustration of TAM (TM Forum 2015f, pp. 22, 63)

The relevant reference process flows are supported through the TAM domains (level 0) *Market/Sales Management* and *Customer Management* (TM Forum 2015e, pp. 22, 63). A more detailed mapping between the reference process flows and TAM application areas (level 1) is provided in Table 4.4. Most application areas are used in several reference process flows, but with different topical focus: for example, Customer Order Management is used in the reference process flows Order-to-Payment and Request-to-Change. The first requires the processing of customer orders to deliver a product as part of a sales scenario. The second deals with changes of a delivered product, which is realized through a change order.

The Customer-Centric domain deals with different types of customer contacts, which could be either related to orders, problems, or complaints. The TM Forum reference model TAM defines these functions in the TAM domain (level 0) *Customer Management* as follows (TM Forum 2015e, p. 63):

- *Customer Information Management* (level 1) defines the management of customer profiles, interrelation between customers and customer groups, the collection of interactions with the customer, subscribed products of the customer, and information about his credit. Those functions are required in the customer touch points across all channels. A standardization of these functionalities for all channels is recommended.
- *Customer Order Management* (level 1) covers all functions to process customer orders during the whole customer order lifecycle. Major functions are the creation of a customer order and its distribution, orchestration, and tracking. Those functions are required in the customer touch points, which should act as single point of contact to the customer. The execution of customer orders requires interfaces to the production (technology domain supported by OSS).
- *Customer Self Management* (level 1) enables the customer to process certain use cases through a self-service portal. Those use cases are structured into fulfillment, assurance, and billing. Self-service portals are normally related to the Internet channel. In addition, they could be realized through automated functions in the call center. Self-service functions vary according to their level of automation. They could be completely automated (e.g., customer profile information) or semi-automated (e.g., customer request is created in a self-service portal and forwarded to a back office agent). In both cases, the Customer Self Management requires interfaces to further functions.
- *Customer Problem Management* (level 1) covers all functions to handle customer problems. Major functions are the creation and qualification of a reported customer problem, conducting a problem diagnosis, as well as the resolution, verification, and closure of a customer problem. The management of customer problems is structured in a hierarchical manner. The Customer Problem Management provides functions for a 1st level support, which might require further support by the service and resource problem management, located in the technology domain.
- *Billing Inquiry, Dispute and Adjustment Management* (level 1) provides functions to deal with billing complaints. They are structured into functions for billing inquiries, disputes, and adjustments. These functions are related to the complaint process, which should be standardized across the different contact channels.

The billing of delivered products is part of the customer-centric domain. The required functions are also included in the TAM domain (level 0) *Customer Management* and structured as follows (TM Forum 2015e, p. 63):

- *Bill Calculation* (level 1) covers the calculation and generation of a customer invoice. Functionalities include the consideration of discounts and tax. The bill calculation can be organized in billing cycles.
- *Receivables Management* (level 1) covers functions to deal with accounts receivable. Those functions include financial activities such as balancing of accounts and financial reporting. Relevant for the customer-centric domain is the payment management. Which includes payment interfaces, validation, authorization, and settlement.
- *Charge Calculation and Balance Management* (level 1) provides functions for customer-specific charges and balances.
- *Collection Management* (level 1) includes functions for the definition and execution of collection policies as well as the concrete settlement of collections. Relevant for the customer-centric domain are the execution of collection policies and the collection settlement.

The availability of all required information at the customer touch points and the management of contact functionalities are requirements of the customer-centric domain that are supported by the following application areas of the TAM domain (level 0) *Customer Management* (TM Forum 2015e, p. 63):

- *Customer Service Representative Toolbox* (level 1) provides a summarized and standardized view on various tools across different channels. Those tools are structured into fulfillment, assurance, and billing applications. The toolbox provides a central access to these functions, which are realized by other application areas: as an example, the fulfillment application contains an order capture function which is linked to the Customer Order Management.
- *Customer and Network Care* (level 1) contains various functions for the management of contact centers—e.g., center administration. Relevant for the customer-centric domain are functions that are related to the contact channel management—e.g., voice channel contact.

Based on the concrete specifics of the customer-centric domain, further functions of the TAM domain (level 0) *Market/Sales Management* might be relevant. These functions are related to sales channels and portals, contract management, and the management of solution-based products. Those application areas include both strategic functions (e.g., forecast analysis) and operational functions (e.g., order tracking capabilities). For the customer-centric domain, the operational functions of the following TAM application areas are relevant (TM Forum 2015e, pp. 22, 63):

- *Channel Sales Management* (level 1) includes functions for the planning and support of different sales channels. Different roles are supported, such as the sales representative and sales administrator. In the customer-centric domain, those functions are realized through an interface to the Customer Management.

- *Sales Portals* (level 1) are divided into internal and indirect sales portals. They cluster specific functions that are relevant for a concrete sales group or channel. Those functions are realized through interfaces to other application areas: for example, an indirect sales portal might offer a customer order creation function for a concrete indirect sales agent which is linked to the customer order management for realizing this function.
- *Contract Management* (level 1) covers functions for the generation, implementation, tracking, and storage of contracts. Customer-specific contracts are typically related to solution-based products for corporate customers. The delivery of the product is part of the customer order, which is included in Customer Management.
- *Solution Management* (level 1) contains functions that are required to provide an offer for a solution-based product. This includes the design and pricing of the solution as well as the offer management and negotiations. Those functions are mainly related to corporate customers.

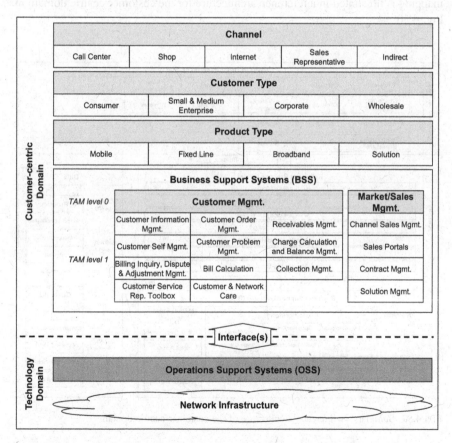

Fig. 4.49 Summary of application layer of customer-centric domain

The relevant functions of the customer-centric domain are summarized in Fig. 4.49. Those functions are related to the BSS and are realized in concrete software systems, for example, in a CRM-software system. The realization of these functions is influenced by the specifics of the customer-centric domain, which can be structured across the dimensions *channel*, *customer*, and *product*. It is recommended to strive for standardization across these dimensions. In addition, a standardized interface between the BSS and OSS is advisable. For instance, in a historically grown implementation, the software systems for customer order management might vary for different product types, which could create additional effort for the sales person—e.g., in the call center.

4.3.5   Summary of the Customer-Centric Domain

The above described content of processes, application, data, and organizational mapping is illustrated in a reference architecture for the customer-centric domain in Fig. 4.50.

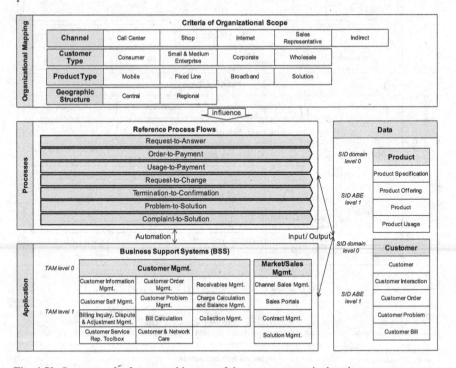

Fig. 4.50 Summary of reference architecture of the customer-centric domain

The customer-Centric domain deals with customer interactions that are initiated by the customer. Those interactions can be summarized in seven use cases: the answering of requests, selling of products, usage of products, change of existing products/contracts, termination of existing products/contracts, reporting of technical problems, and reporting of commercial complaints. These seven use cases are reflected in the reference process flows which are automated by Business Support Systems (BSS). Based on the TM Forum reference model TAM (TM Forum 2015e, pp. 22, 63), the functions of the BSS can be summarized in *customer management* and *market/sales management*. Both processes and applications use the data entities *product* and *customer*, which are defined according to the TM Forum reference model SID (TM Forum 2015b, pp. 19–24).

In a concrete implementation, the detailed design of processes, applications, and data depends on the organizational scope, which can be summarized based on the dimensions channel, customer type, product type, and geographic structure. It is recommended to standardize the design across these dimensions. Based on the reference process flows a high level of standardization should be targeted. This is a prerequisite for the aggregation of application functions in a small number of integrated software systems, which is supported by a central data model.

The separation between customer view and production view is essential. The customer-centric domain focuses on products and customers which are realized in the technology domain through services, resources, and subscribers. The objective is a decoupling from the technical complexity. The concrete network technologies and their realization through network elements should be a black box for the customer-centric domain. Hence, standardized interfaces between customer-centric and technology domains are essential.

4.4   Detailing the Technology Domain

The technology domain covers the roll-out, extension, operations, and maintenance of the network infrastructure as well as the development and realization of telecommunication services. The reference process flows (cf. Sect. 4.4.1) provide a recommendation for the process layer. The mapping between the technology domain and the organizational structure is an essential step in a real-life transformation project (cf. Sect. 4.4.2). Furthermore, recommendations for the data layer (cf. Sect. 4.4.3) and application layer (cf. Sect. 4.4.4) are provided. The interrelations between those different layers are summarized in high-level illustrations of the technology domain (cf. Sect. 4.4.5).

4.4.1   Reference Process Flows of the Technology Domain

The *technology domain* includes all reference process flows[12] related to development, provisioning, operations, and maintenance of services and resources (cf. Fig. 4.51). It provides the technical view to the customer-centric and product domains.

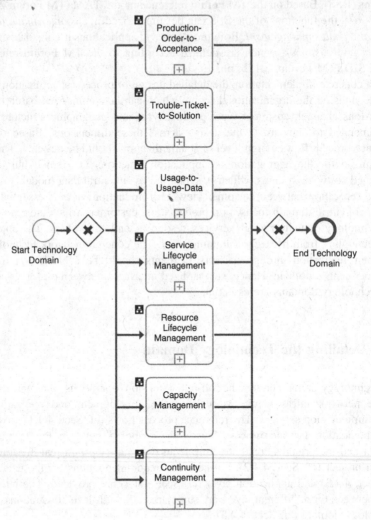

Fig. 4.51 Reference process flows of technology domain (level 1)

---

[12]The reference process flows are based on intensive project work, research and development. This work was mainly conducted by Detecon in various different teams. The two authors had leading roles in this development. The reference processes flows of the technology domain were already published in prior scientific publications (e.g. Czarnecki 2013; Czarnecki et al. 2013; Czarnecki and Spiliopoulou 2012) and white papers (e.g. TM Forum 2015e, 2010). The descriptions and illustrations in this section are a completely revised version.

Production-Order-to-Acceptance (cf. Fig. 4.52) receives a production order[13] and covers its execution. It ends with the acceptance of the executed production order. The production order can be received from the customer-centric domain (e.g., provisioning of a product), the product domain (e.g., realization of a new product), or the technology domain (e.g., realization of a service). Based on the production order, the required services and resources are identified, comparable with the explosion of a bill of material. Then the relevant tasks are scheduled and executed, which might result in the issuing of further production orders.

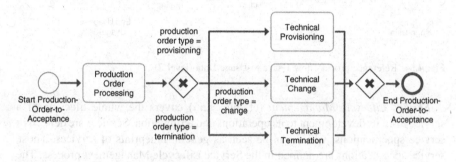

Fig. 4.52 Reference process flow Production-Order-to-Acceptance (level 2)

Trouble-Ticket-to-Solution (cf. Fig. 4.53) covers the maintenance of services and resources. It starts with the receiving of a trouble ticket that reports a technical problem. This trouble ticket might have different origins. If technical problems reported by the customer cannot be solved by the customer-centric domain, they are forwarded as trouble tickets to the technology domain (e.g., as a 2nd level support). During the operations of services and resources, technical problems are automatically detected (so-called alarms). Those alarms might result in a trouble ticket (e.g., based on the impact of the alarm). The Trouble-Ticket-to-Solution process includes all technical activities that are required to solve the problem. In most cases, it is structured in different levels and might include an interface to suppliers.

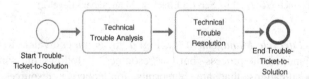

Fig. 4.53 Reference process flow Trouble-Ticket-to-Solution (level 2)

---

[13]The reference data model SID (TM Forum 2015b) does not differentiate between production orders and work orders. Therefore, production orders are realized through work orders in the data model.

Usage-to-Usage-Data (cf. Fig. 4.54) provides the technical view to the Usage-to-Payment process of the customer-centric domain—i.e., it covers the usage of services and resources based on subscriptions. Those subscriptions are related to contracts, customers, and products in the customer-centric domain. The process starts with the authorization of a usage request. It collects the required usage data, including the rating according to the defined subscription (e.g., for a usage-based tariff).

Fig. 4.54 Reference process flow Usage-to-Usage-Data (level 2)

Service Lifecycle Management (cf. Fig. 4.55) covers the whole lifecycle of a service from its development to its operations and termination. Services are defined in service specifications, which can be seen as general blueprints of services. Those service specifications are defined in the Service Lifecycle Management process. The provisioning of a concrete service—for example, as part of production order received from the customer-centric domain—is realized through an instantiation of the service specification in the Production-Order-to-Acceptance process. A new service specification can be either triggered by a product requirement from the product domain or by a technical requirement from the technology domain itself. This new service specification can be either realized through existing resource specifications or result in a new requirement for the Resource Lifecycle Management process.

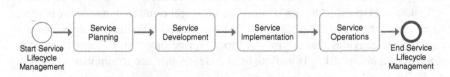

Fig. 4.55 Reference process flow Service Lifecycle Management (level 2)

Resource Lifecycle Management (cf. Fig. 4.56) is the equivalent to the Service Lifecycle Management process, but for resources. It follows the same logic of resource specifications that are blueprints for concrete resources. However, resources can be related to physical elements of the technical infrastructure. New resources or resource specifications might require physical work. Most infrastructure elements are provided by external suppliers with different specifics and limitations. The aim of the Resource Lifecycle Management is an overall management of all resources.

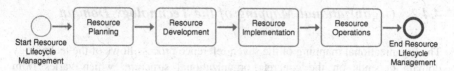

Fig. 4.56 Reference process flow Resource Lifecycle Management (level 2)

Capacity Management (cf. Fig. 4.57) deals with the capacity of the technical infrastructure. It covers the whole lifecycle: the monitoring and evaluation of existing capacities, the identification of capacity shortages, the definition of capacity targets, and the implementation of new capacities. Capacities are important for both the commercial and the technical view. From the commercial view, they are directly related to the satisfaction of market needs and usage limitations for the customer. From a technical view, capacities are required for proper operations of services and resources.

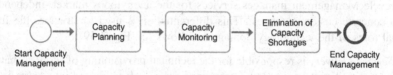

Fig. 4.57 Reference process flow Capacity Management (level 2)

Continuity Management (cf. Fig. 4.58) aims for a high reliability of the technical infrastructure as well as recovery mechanism in case of failures. The process covers all related activities, from the definition of continuity strategies and plans to their implementation. It includes such issues as emergency plans as well as planning of infrastructure redundancies. Requirements for the Continuity Management process can be received from various other processes, such as Corporate Strategy and Management (e.g., budget restrictions), Legal and Regulatory Management (e.g., legal obligations), and Business-Opportunity-to-Launch (e.g., service level agreements related to a new product).

Fig. 4.58 Reference process flow Continuity Management (level 2)

4.4.2 Organizational Mapping of the Technology Domain

The organizational mapping of the seven reference process flows of the technology domain depends on the concrete organizational structure, which varies from enterprise to enterprise. Besides the organizational specifics of the enterprise, it is also influenced by the technical details of the network infrastructure. Similar to the customer-centric domain, the mapping of the technology domain is supported by a description of general functional roles. These general functional roles are derived from the experience with various implementation projects. They follow a functional organizational structure which is orthogonal to the process perspective.

The reference process flows of the technology domain can be divided into customer-facing and non-customer-facing. The customer-facing reference process flows interact with the customer-centric domain in order to technically fulfill a customer request. For example, Production-Order-to-Acceptance receives the request from the customer-centric domain to provision a product based on a customer order. In contrast, the non-customer-facing reference process flow Service Lifecycle Management manages services for the anonymous market, independent of a concrete customer request. This differentiation is used to structure the functional roles of the technology domain as follows (cf. Fig. 4.59):

- *Service Delivery* is responsible for the technical provisioning of services that are related to a customer order. A customer order is decomposed into work orders that are executed by the service delivery. Customer orders can be related to sales, change, or termination requests. For consumer customers especially, most of the service delivery is automated. The field service is a typical organizational entity that executes physical work of the service delivery. The service delivery belongs to the customer-facing part of the technology domain.

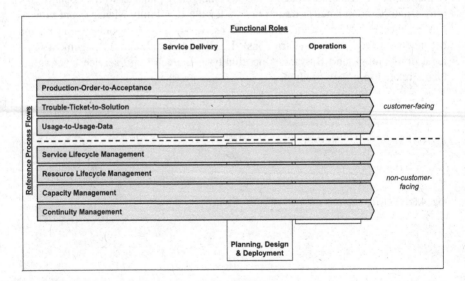

Fig. 4.59 Functional roles involved in technology domain

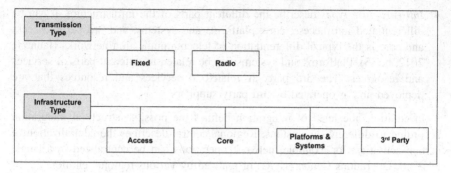

Fig. 4.60 Criteria of organizational scope in the technology domain

- *Planning, Design and Deployment* covers all activities that are related to the development and realization of new services, the change of existing services, the change of existing network resources, and the roll-out of new infrastructure. All of these activities are irrespective of concrete customer orders, and are therefore non-customer-facing.[14] They can be either related to new product requirements received from the product domain, or related to purely technical requirements from the technology domain itself.
- *Operations* is responsible for all activities related to the operations of service and resources. After the roll-out of services and resources, they are handed over to operations. Their activities are either customer-facing (e.g., 3rd level support for a customer problem), or non-customer-facing (e.g., regular technical maintenance cycles).

In the past, the deployment and operations of network infrastructure were an indispensable prerequisite for a telecommunications operator. Therefore, their organizations were arranged around the technical specifics of the network influenced by their network suppliers (Lewis 2001, p. 242; Misra 2004, p. 3). Today, a telecommunications operator does not necessarily require an own network infrastructure but can outsource a variable part of it. Hence, the definition of organizational entities depends heavily on the scope of the concrete technical infrastructure, which can be structured as follows (cf. Fig. 4.60):

- *Transmission types* are typically separated into radio and fixed. Both transmission types have different requirements, which are in most cases reflected in the organizational structure.

---

[14]The delivery of solution-based products might require the extension of network infrastructure based on a customer order (especially for corporate customers). This case is included in the functional role of service delivery.

- *Infrastructure types* describe the different parts of the infrastructure. It can be differentiated into access, core, platforms and systems, and 3rd party. Access and core is the typical differentiation of telecommunication networks (Iannone 2012, p. 35). Platforms and systems can be related to different parts of services and resources. The 3rd party is related to services and resources that are deployed and/or operated by 3rd party suppliers.

In addition, the level of integration defines the possible structural integration according to different dimensions. Regional/central describes the centralization of certain activities; for example, network operations can be centralized in a single Network Operations Center (NOC) or realized by various regional entities.

According to the characteristics of the above criteria, the relevant functional roles of service delivery, operations, and planning, design and deployment are defined in different ways. A description of exemplary specifications of organizational designs is given below.

Typically, telecommunication operators have a strict organizational separation between network planning and network operations (Hämäläinen et al. 2012, p. 6). Figure 4.61 shows an exemplary organizational structure that distinguishes on a high level between network planning and design, network deployment and network operations. An important concept of the TM Forum reference models is the differentiation between service and resource (e.g. TM Forum 2015b, pp. 24–31). In this example, the term *network* is related to services and resources. A differentiation between both could be realized in the detailed organizational structure, e.g., Service Planning and Design.

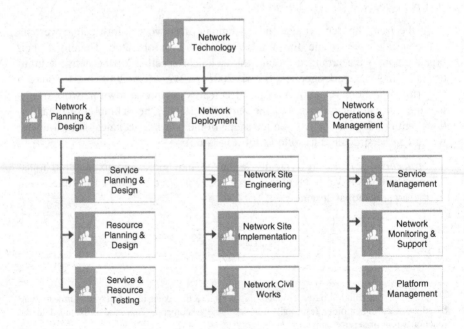

Fig. 4.61 Exemplary organizational structure distinguishes between functions

Network operations works on the live network (Hämäläinen et al. 2012, p. 6). Both the management and operations of the network are important prerequisites for a proper delivery and usage of products. It is continuously involved into the daily business of telecommunications operators. Figure 4.62 shows an exemplary organizational structure for network operations and management. The service management is responsible for the service delivery as well as the management of service quality and performance. The network monitoring and support continuously monitors the network and reacts to network problems. Those problems could be either based on alarms automatically reported by the network or problems reported by customers. Based on the problem type, the 1st level support, 2nd level support, or field service is required for the problem resolution. For customer reported problems, there is an interface to the customer contact channel (e.g.. call center). The single point of contact should remain in the customer-centric domain. Platform management is responsible for configuration, administration, and maintenance of the network platforms. The tasks of the network operations and management organization are typically realized in a Network Operations Center (NOC) (Mishra 2007, p. 461).

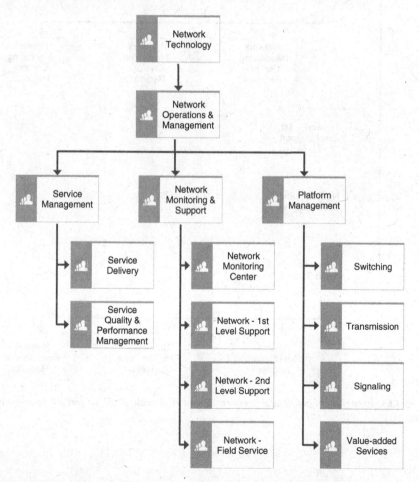

Fig. 4.62 Exemplary organizational structure of network operations and management

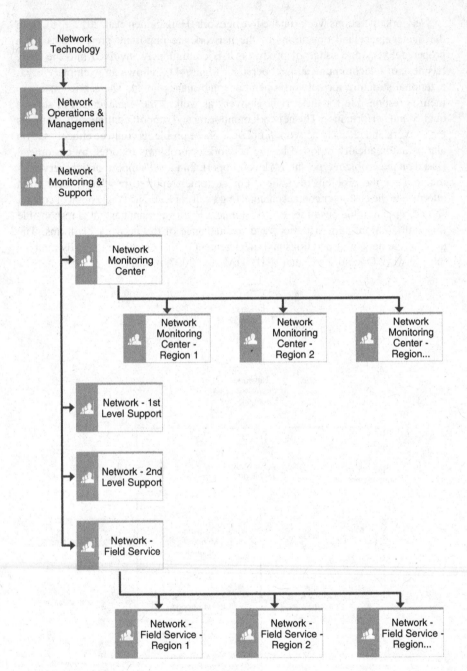

Fig. 4.63 Exemplary organizational structure of network operations and support differentiated by regions

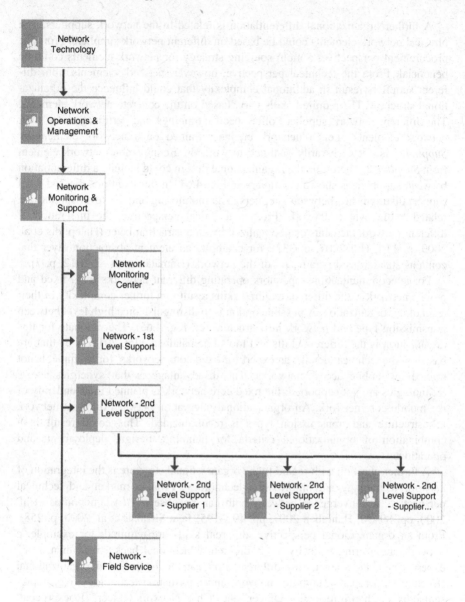

Fig. 4.64 Exemplary organizational structure distinguishes between network element suppliers

Based on the geographic structure of the telecommunications operator, the physical structure of the network influences the organizational structure. In particular, the network operations as well as the network deployment might require local organizational entities. Figure 4.63 shows an exemplary differentiation of the network monitoring and field service by regional entities.

A further organizational differentiation is related to the network suppliers. The physical network elements could be based on different network suppliers. From the procurement perspective, a multi-sourcing strategy for network elements could be beneficial. From the technical perspective, however, network elements from different suppliers result in additional complexity that could influence the organizational structure. The required skills vary, based on the concrete network elements. The different network suppliers offer specific trainings and certification for their network elements—i.e., a network engineer trained on a network element by *Supplier 1* is not necessarily qualified to work on a comparable network element from *Supplier 2*. Therefore, the organizational design could include a differentiation between suppliers as shown exemplary in Fig. 4.64. In this example, the 2nd level support distinguishes between suppliers. The monitoring and 1st level support are related to the whole network. From a technical perspective, the integration of different network elements can be realized by an abstraction layer (Haleplidis et al. 2009, p. 110; Hill 2007, p. 332), for example, an element abstraction layer that contains standardized parameters of the network (Hämäläinen et al. 2012, p. 7).

For telecommunications operators operating different networks (e.g., fixed and mobile networks), the different required skills result in a further complexity of their technology organization. A possible option is to distinguish on a high level between transmission type and network infrastructure (cf. Fig. 4.65). The rationale for this organization is the different skills and tools (Hämäläinen et al. 2012, p. 7) that are based on the technical details; an expert for fixed core networks, for example, is not trained on mobile access networks. The disadvantage is that synergies across technologies are not supported; the fixed core network is planned independently of the mobile core network. An organizational integration between different network infrastructure and transmission types is recommended. This could result in a combination of organizational criteria for planning, design, deployment, and operations.

A further question with regard to the organizational structure is the integration of network technology and information technology. From a market and technical perspective, a convergence between both can be observed (Adamopoulos et al. 2000, pp. 89–90; Hanrahan 2007, pp. 194–195; Jaya Shankar et al. 2000, p. 258). From an organizational perspective, different skills are required; for example, a network engineering typically has a different educational background than an IT expert. On a high level, this differentiation can be observed in organizational structure with separate units for network and information technology, e.g., represented by a "Chief Information Officer" and "Chief Network Officer" (Rockart et al. 2003, p. 308). A first integration could result in a "Chief Technology Officer" heading two separate departments responsible for networks and information technology (cf. Fig. 4.66). Still, there are reasons for a further integration on a lower organizational level. Software systems used to managing and operating network infrastructure are highly dependent. There is an on-going convergence between

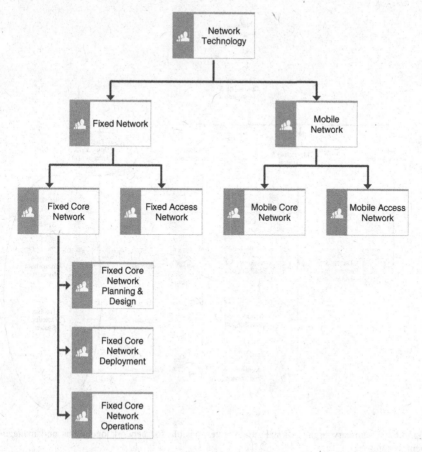

Fig. 4.65 Exemplary organizational structure distinguishes between transmission types and network infrastructure

network and software resources that are both combined in an end-user product (Yahia et al. 2006). Network and IT virtualization result in an integrated operation of network and IT infrastructure (Buhl and Winter 2008, p. 135; Kusnetzky 2011, pp. 23–27). Figure 4.67 shows an exemplary integration of organizational entities responsible for systems that are related to network operations and management.

Fig. 4.66 Exemplary high-level organizational integration of network and information technology

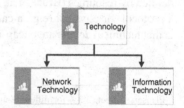

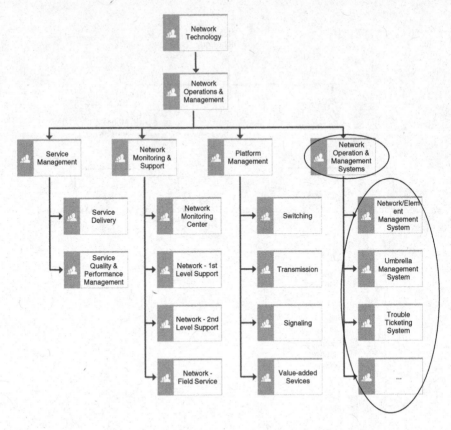

Fig. 4.67 Exemplary organizational structure responsible for network operations and management systems

4.4.3 Data Layer of the Technology Domain

The data processed in the technology domain is mainly related to the TM Forum SID domains *service* and *resource* (TM Forum 2015b, pp. 24–31). Resources are required to realize services which are then assembled into products (Bruce et al. 2008, p. 19; Snoeck and Michiels 2002, p. 335). In the meaning of SID, services are functional components of a product, which could be tangible (e.g., a mobile phone) or intangible (e.g., a broadband connection).[15] The concrete technical realization is defined by resources (Bruce et al. 2008, p. 19). Resources can be either logical (e.g., a protocol) or physical (e.g., a cable modem). The advantage of this differentiation is that technical details are encapsulated in the resources. This allows abstracting the

---

[15]Please see Sect. 3.3.4 for differentiation between the term *service* in SID and ITIL.

product as well as the customer from technical complexities. This differentiation between service and resource is comparable to the separation between service and transport that is used in a Next Generation Network (e.g. Knightson et al. 2005, p. 49).

Services are connecting products that are sold to customers with resources that are required for their realization. There is a relation between services and products as well as services and resources. Both relations are defined from completely different perspectives. The relation between service and product is defined from the business perspective and is abstracted from technical details. The relation between service and resource is defined from the technical perspective and requires technical details. In order to combine both perspectives in the service entity, the reference data model SID differentiates between *customer-facing services* and *resource-facing services* (TM Forum 2015b, p. 46). *Customer-facing services* are directly linked to one or more product(s). This relation is a n:m-relation—i.e., a customer-facing service can be linked to various different products and a product can be comprised of different customer-facing services. Resource-facing services are linked to resources. In addition, there is a connection between customer-facing services and resource-facing services.

With the above described structure, the provisioning of a concrete product sold to a customer can be defined by customer-facing services linked to resource-facing services which are realized by resources. Furthermore, a differentiation between a concrete product instance and its definition is required. A product instance is owned by a customer and linked to concrete physical resources; some examples could be a mobile phone in the customer's possession, or a clearly located mobile base station used by the customer for a mobile phone connection. The definition of a product is a blueprint that provides a general composition; for example, a mobile tariff consists of a selection of different mobile phone models and the possibility to use a mobile network. The reference data model SID terms the definition of a product and its components as *product specification*, *service specification*, and *resource specification* (TM Forum 2015b, p. 46). Further details on the product development are described in Sect. 4.5.

In summary, the definition of products and their technical realization are defined in flexible data structures. In an ideal scenario, those specifics would be completely decoupled from processes and applications. A new product specification would be defined through a new data relation between the product specification and customer-facing service specification(s). Each customer-facing service specification is realized through a new or existing relation with resource-facing service specification(s). Those relations are defined in the reference process flows *service lifecycle management* and *resource lifecycle management* by using the functions of the application layer, such as the *service inventory* and *resource inventory*. In an ideal scenario, new products are also realized through data changes based on standardized processes and applications.

For both product development and provisioning, automated and manual work might be required. In the product domain, the product development is structured by projects which are decomposed into work orders. Those work orders are executed in the technology domain. The provisioning is initiated by a customer order that is realized by work orders in the technology domain.[16] The technical realization in the technology domain is in both cases organized through work orders.

After a successful provisioning, a product is operated by the technology domain and used by the customer. During this usage, the relevant usage data is collected— for example, for the bill calculation. The usage data is derived from service usage that is based on resource usage. In case of technical problems reported by the customer, a customer problem is created. This customer problem can be linked to a service problem, which could be related to a resource trouble. Customer problems are handled in the customer-centric domain. Service problems and resource troubles belong to the technology domain. In addition, the continuous monitoring of the services and resources might detect an alarm which is not related to a concrete customer problem. In this case, the service problem or resource trouble is not necessarily linked to a customer problem. Furthermore, a continuous monitoring of the service and resource performance is conducted by the technology domain. The results are consolidated into the product performance.

The interrelation between the SID data entities of the product domain, customer-centric domain, and technology domain for product development, provisioning and usage is illustrated in Fig. 4.68.

The reference model SID contains detailed entity relationship models for all described data entities. In the following is a short summary of the major Aggregated Business Entities (ABE) that are relevant for the technology domain (TM Forum 2015b, pp. 24–31):

- *Service Specification* (ABE level 1) and *Resource Specification* (ABE level 1) define the characteristics of services and resources. Both are components of product specifications. They are structured in a hierarchical manner.
- *Service* (ABE level 1) and *Resource* (ABE level 1) are concrete instances of a service or resource specification.
- *Service Problem* (ABE level 1) and *Resource Trouble* (ABE level 1) are related to technical problems either from a service or resource perspective. Both could be either based on a reported customer problem or a monitored alarm.
- *Service Usage* (ABE level 1) and *Resource Usage* (ABE level 1) contains usage data related to services or resources. Those statistics could be summarized and forwarded to the customer-centric domain, e.g., for billing purposes.
- *Service Performance* (ABE level 1) and *Resource Performance* (ABE level 1) contains performance statistics either from a service or resource perspective.

---

[16]From a functional perspective TAM uses the terms *customer order*, *service order*, and *resource order* to describe the provisioning of a product. From a data perspective *service orders* and *resource orders* are summarized in *work orders* that are executed in the technology domain.

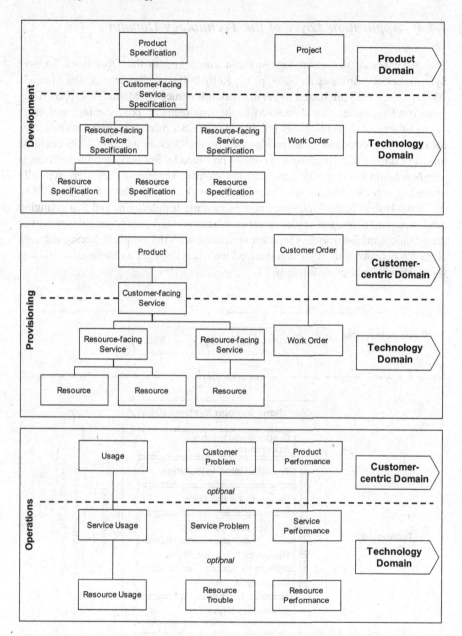

Fig. 4.68 Usage of SID data entities for product development, provisioning, and usage

4.4.4 Application Layer of the Technology Domain

The processes of the technology domain are linked to the Operations Support Systems (OSS) (Bruce et al. 2008, p. 15; Kelly 2003, p. 109; Snoeck and Michiels 2002, p. 331). OSS are linked to the network infrastructure (Misra 2004, p. 3). They support all processes that are related to the management, provisioning, and operations of services and resources (cf. Fig. 4.69). OSS receive requests from BSS—for example, a customer order provisioned by the OSS. In addition, OSS could be linked to further infrastructure and content provided by 3rd party suppliers. There is a tight relation between OSS and network element suppliers. They were typically offered as supplier-specific solutions for their own network elements (Misra 2004, pp. 3–4). Due to today's complex communication technologies and fast changing market requirements, the objective are standardized OSS which are independent of the technical infrastructure. The communication with the supplier-specific network elements is realized through standardized interface(s)—e.g., an element abstraction layer (Hämäläinen et al. 2012, p. 7).

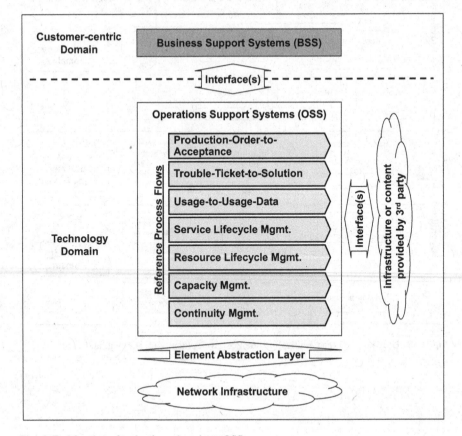

Fig. 4.69 Mapping of technology domain to OSS

According to Misra (2004) the following major functions of OSS can be differentiated:

- *Element management* provides basic functions on the network resources, like e.g. reporting of technical alarms as well as execution of operational commands.
- *Traffic management* covers the monitoring of network traffic as well as the generation of traffic statistics. The traffic management can be related to different aggregation levels.
- *Configuration management* includes the installation of network elements, implementations related to path and capacity, as well as definition of network protection parameters.
- *Service provisioning and activation* realizes the required changes and configuration of service and resources in order to fulfill a customer request.
- *Fault management* covers network monitoring, fault diagnostics, and network testing. It is typically related to the Network Operations Center (NOC).
- *Administration and security* includes the restriction of the usage and access to network related systems and resources.
- *Network planning support* covers the planning and design of network resources. It contains functions for evaluation and simulation of network capacities.
- *Inventory management* stores information about service and resources.

Based on the reference process flows of the technology domain, the relevant TAM domains (level 0) are *Service Management* and *Resource Management* (TM Forum 2015e, pp. 171, 213). A more detailed mapping between the reference process flows and TAM application areas (level 1) is provided in Table 4.5.

The technology domain receives all customer orders from the customer-centric domains which could be related to the provisioning of new products, the change of existing products, and the termination of existing products. Those customer orders are related to production orders that are executed by the Production-Order-to-Acceptance process of the technology domain. The production order is decomposed into a provisioning of services and resources. In this respect, a production order could be understood as a collection of service orders and resource orders.[17] The TM Forum reference model TAM defines the related functions as follows (TM Forum 2015e, pp. 171, 213):

- *Service Order Management* (level 1) covers all functions required to process service orders. Major functions are the assessment of the service availability, the decomposition of customer orders into service orders, the decomposition of service orders into resource orders, the tracking of service orders, and the configuration and activation of services.

---

[17]From a functional perspective, TAM uses the terms *service order* and *resource order* to describe the processing of a production order on the service and resource level. From a data perspective, service orders and *resource orders* are summarized in *work orders* that are executed in the technology domain.

Table 4.5 Mapping between reference process flows and application areas (level 1) for technology domain

Reference Process Flows	Application Areas (Level 1) (TM Forum 2015f, pp. 171, 213)[a]
Production-Order-to-Acceptance	Service order management, resource order management, workforce management, service test management, resource test management
Trouble-Ticket-to-Solution	Service problem management, fault management, workforce management, service test management, resource test management
Usage-to-Usage-Data	Usage management
Service Lifecycle Management	Service inventory management, service performance management, service quality management, service test management
Resource Lifecycle Management	Resource lifecycle management, resource inventory management, resource performance management, workforce management, resource test management
Capacity Management	Resource lifecycle management, resource performance management, workforce management
Continuity Management	Resource lifecycle management, workforce management

[a]The mapping between the reference process flows and application areas is an own proposal. The application areas are based on the high-level illustration of TAM (TM Forum 2015f, pp. 171, 213)

- *Resource Order Management* (level 1) provides similar functions like the Service Order Management, but for resources. Major functions are the validation of resource orders, the tracking of resource orders, and the configuration of resources.

The resolution of technical problems is included in the Trouble-Ticket-to-Solution process of the technology domain. Those problems are either forwarded by the customer-centric domain or identified through alarms in the technology domain. The TM Forum reference model TAM defines the related functions as follows (TM Forum 2015e, pp. 171, 213):

- *Service Problem Management* (level 1) covers the receiving, monitoring, analysis, resolution, and tracking of service problems. Even though service problems can be related to customer problems, in most cases a consolidated view is required in order to identify and eliminate the root cause.
- *Fault Management* (level 1) is related to technical resource problems. It covers functions for monitoring, root cause analysis, correction, and restoration of resources.

The whole lifecycle of services and resources is managed in the technology domain. The objective is to assure a timely fulfillment of customer request through the continuous planning, design, and implementation of services and resources. The TM Forum reference model TAM defines the related functions as follows (TM Forum 2015e, pp. 171, 213):

- *Service Inventory Management* (level 1) and *Resource Inventory Management* (level 1) store the information about services and resources. This contains the logical interrelation between services and resources as well as the physical appearance of resources. The service and resource inventories are an important prerequisite for process standardization because they contain all network-specific information in a central repository.
- *Service Performance Management* (level 1) and *Resource Performance Management* (level 1) provide functions for monitoring, analysis, and reporting of service and resource performance. The results of the Resource Performance Management are an input for the Service Performance Management that consolidates the resource performance on a service level.
- *Service Quality Management* (level 1) covers the definition, monitoring, analysis, and reporting of service quality.
- *Resource Lifecycle Management* (level 1) supports the planning, design, specification, configuration, and implementation of resources. Those functions are independent of customer orders, but related to continuous planning and extension activities of the network infrastructure.

In addition, the TM Forum reference model TAM contains various functions that are related to the whole technology domain (TM Forum 2015e, pp. 171, 213):

- *Usage Management* (level 1) provides usage data related to network resources. The usage data is composed based on usage events and forwarded to the relevant recipient functions, e.g., Bill Calculation.
- *Workforce Management* (level 1) manages the field service and related resources (e.g., vehicles). Based on the technical details, the field service could be involved in various parts of the technology domain—e.g., service provisioning, resource problems, or resource implementations.
- *Service Test Management* (level 1) and *Resource Test Management* (level 1) provide functions for the testing of services and resources. They are involved in the fulfillment and assurance, which could be related to a concrete customer order as well as a customer independent network maintenance or implementation.

The relevant functions of the technology domain are summarized in Fig. 4.70. These functions are related to the OSS and realized in concrete software systems. Those software systems are either offered by suppliers of network elements or irrespective of network suppliers. The objective is supplier and technology-independent OSS, which could be realized by an element abstraction layer. The concrete realization of these functions is influenced by the specifics of the technology domain which can be structured across the dimensions *transmission*

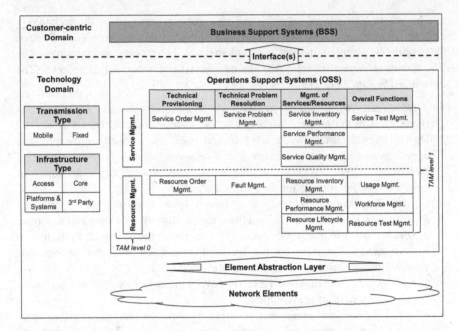

Fig. 4.70 Summary of application layer of technology domain

type and *infrastructure type*. The functions can be summarized in technical provisioning, technical problem resolution, management of services/resources, and overall functions. Those functions are structured into services and resources. The service-related functions provide a generalization of technical components which are independent of the technical realization. In contrast, the resource-related functions are related to concrete physical or logical elements. This decoupling of services and resources allows an encapsulation of technical specifics in small parts of the application layer. It is recommended to strive for a standardized interface between the OSS and BSS, as well as the OSS and network elements.

4.4.5 Summary of the Technology Domain

The above-described content of processes, applications, data, and organizational mapping is illustrated in a reference architecture for the technology domain in Fig. 4.71.

The technology domain is responsible for the development, provisioning, and operations of services and resources. Those services and resources are the components that are assembled into products. Resources are related to the physical infrastructure (e.g., network elements). Services are technical functionalities that are independent of their concrete technical realization. This differentiation between

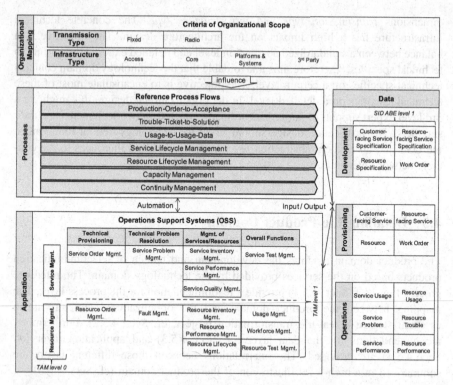

Fig. 4.71 Summary of reference architecture of the technology domain

service and resource is a major concept of the technology domain. Furthermore, the concrete instance of a service or resource is distinguished from its specification. A service or resource specification is a blueprint that is used during the provisioning in order to create a concrete instance of a service or resource.

The activities of the technology domain are summarized in the seven reference process flows. The first three reference process flows Production-Order-to-Acceptance, Trouble-Ticket-to-Solution, and Usage-to-Usage-Data are related to customer requests in the customer-centric domain. The other four reference process flows are responsible for the development and operations of the technical infrastructure. All these activities are automated in the Operations Support Systems (OSS). They can be structured into technical provisioning, technical problem resolution, management of services/resources, and overall functions. Based on the TM Forum reference model TAM (TM Forum 2015e, pp. 171, 213), the functions of the OSS are related to the TAM domains *service management* and *resource management*. Both processes and applications use the data entities *service* and *resource*, which are defined according to the TM Forum reference model SID (TM Forum 2015b, pp. 24–31).

In a concrete implementation, the detailed design of processes, applications, and data depends on the organizational scope, which can be summarized based on the dimensions *transmission type* and *infrastructure type*. The concrete technical infrastructure has a high impact on the architecture design. The objective is a balance between a standardized abstraction level and detailed expertise according to technical specifics. Therefore, it is unavoidable that the technology domain contains technical specifics on a certain level. The objective is to encapsulate most of these technical specifics on a detailed level. In the organizational design, this might be a 3rd level expert support organized according to concrete network elements and their suppliers. The application design should strive for technology-neutral functions in the OSS and an element abstraction layer with standardized interfaces to the concrete network elements.

4.5 Detailing the Product Domain

The product domain contains the development and launch of telecommunication products based on the services provided by the technology domain. The reference process flows (cf. Sect. 4.5.1) provide a recommendation for the process layer. The mapping between the product domain and the organizational structure is an essential step in a real-life transformation project (cf. Sect. 4.5.2). Furthermore, recommendations for the data layer (cf. Sect. 4.5.3) and application layer (cf. Sect. 4.5.4) are provided. The interrelations between those different layers are summarized in a high-level illustration of the product domain (cf. Sect. 4.5.5).

4.5.1 *Reference Process Flows of the Product Domain*

The *product domain* covers the whole product lifecycle from the idea generation for new products to the elimination of outdated products (cf. Fig. 4.72). It is structured into four reference process flows,[18] which are defined by the major events of the product lifecycle. To understand the interrelations with the customer-centric domain, a differentiation between the product specification and a concrete product is important. The product specification is a blueprint for a concrete product sold to a customer. The product domain defines those product specifications, and the customer-centric domain creates instances of them which are concrete products.

Idea-to-Business-Opportunity deals with the generation of new product ideas and the planning of the product portfolio (cf. Fig. 4.73). The idea generation is the first part of this process. However, the appearance of an idea can also be seen as its starting event. Those ideas are evaluated with respect to their commercial and technical

---

[18]The reference process flows are based on intensive project work, research and development. This work was mainly conducted by Detecon in various different teams. The two authors had leading roles in this development. The reference processes flows of the product domain were described on a high level in Czarnecki et al. (2013).

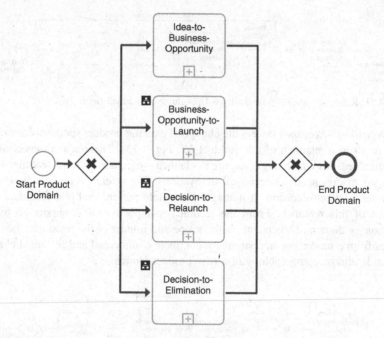

Fig. 4.72 Reference process flows of product domain (level 1)

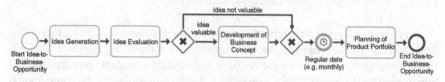

Fig. 4.73 Reference process flow Idea-to-Business-Opportunity (level 2)

impact. The outcomes of this process are concrete business opportunities with a realization decision, required budget, and a realization planning. Those business opportunities are managed in a product portfolio and planned in a product roadmap.

Business-Opportunity-to-Launch is the product launch process. It starts with a concrete business opportunity received from the Idea-to-Business-Opportunity process (cf. Fig. 4.74). The commercial and technical launch of this business opportunity is managed. From the technical perspective, this process is supported by the technology domain. In fact, this process realizes the product specifications which are the blueprints for new products sold by the customer-centric domain. A new product specification could be either realized through existing service specifications or require the realization of new service specifications. After its launch, the new product specification is handed over to the customer-centric domain. A concrete new product sold to a customer is then an instance of this product specification.

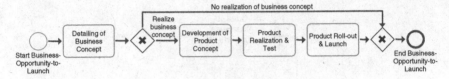

Fig. 4.74 Reference process flow Business-Opportunity-to-Launch (level 2)

Decision-to-Relaunch covers the change of existing product specifications which will result in a relaunch of this product (cf. Fig. 4.75). There are commercial and technical reasons which might require a relaunch—e.g., new product features due to changed market needs or technical innovations. The reference process flow starts with the relaunch decision. It manages the whole commercial and technical realization of this relaunch. From the technical perspective, it is supported by the technology domain. Dependent on the scope and impact of the relaunch, the effort varies from a marketing campaign to a complete commercial and technical change, which is almost comparable with a new product launch.

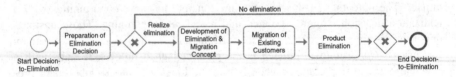

Fig. 4.75 Reference process flow Decision-to-Relaunch (level 2)

Decision-to-Elimination is related to the end of the product lifecycle (cf. Fig. 4.76). The elimination of outdated products is important for complexity reduction. This reference process flow starts with the decision to eliminate a product specification. The commercial and technical impact is evaluated. Based on these results, the elimination is planned. The process ends with the successful product elimination.

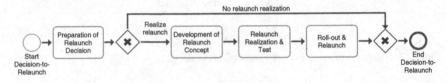

Fig. 4.76 Reference process flow Decision-to-Elimination (level 2)

4.5.2 Organizational Mapping of the Product Domain

The organizational mapping of the four reference process flows of the product domain depends on the concrete organizational structure, which varies from enterprise to enterprise. The product domain translates market requirements into

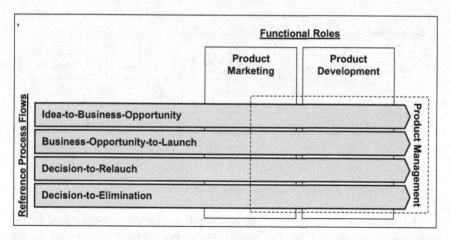

Fig. 4.77 Functional roles involved in product domain

technical requirements that are realized by the technology domain. Those market requirements are independent of concrete customer requests. Once a product is operational, it is sold by the customer-centric domain. Hence, the product domain has interfaces to both technology and customer-centric domains, but it is not involved in the daily business of answering customer requests.

Similar to the other domains, the mapping of the product domain is supported by a description of general functional roles. These general functional roles are derived from the experience with various implementation projects. They follow a functional organizational structure, which is orthogonal to the process perspective.

Major activities of the product domain are the planning, design, development, and roll-out of new products as well as changes to existing products. All these activities are conducted from a business perspective. The technical realization is part of the technology domain. The product domain can be structured into the following functional roles (cf. Fig. 4.77):

- *Product Marketing* is responsible for the analysis of the market and new market requirements, the monitoring and evaluation of product innovations, the generation and evaluation of product ideas, and the overall planning and management of the product portfolio.
- *Product Development* covers all activities that are related to the detailed design, development, and roll-out of products from a business perspective. For the technical realization, the product development works closely together with the technology domain.
- *Product Management* defines the end-to-end responsibility for concrete products. Typically a product manager is nominated for each product in order to manage the whole development, roll-out, and operations. This role is orthogonal to the two roles of product development and marketing.

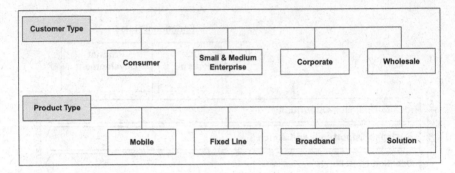

Fig. 4.78 Criteria of organizational scope in the product domain

The organizational structure of the product domain is mainly influenced by the customer segments and product types. These two criteria are similar to the customer-centric domain (cf. Sect. 4.3.2). In the customer-centric domain, these two criteria are related to the operational handling of products with respect to pre-sales, sales, and post-sales. In the product domain, those two criteria are used to structure the marketing, development, and management of products before they are handed over to the customer-centric domain (cf. Fig. 4.78):

- *Customer types* describe the types of customers addressed by the enterprise. They can be differentiated into consumer, small and medium enterprise, corporate, and wholesale.
- *Product types* define the types of products offered by the enterprise. A typical categorization is mobile, fixed line, broadband, and complex solutions.

According to the above criteria, the product marketing and development can be structured based on the customer type and/or product type. Hence, the organizational structure of the product domain might follow a comparable logic to the customer-centric domain. Consumer, corporate, and wholesale markets typically have different conditions and requirements. Therefore, the organizational design for product marketing and development might distinguish between consumer, corporate, and wholesale customers on a high level (cf. Fig. 4.79). This structure is similar to the differentiation of the Sales and Customer Service organization according to customer types (cf. Fig. 4.39).

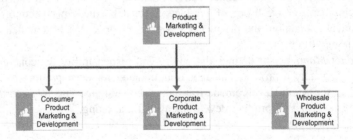

Fig. 4.79 Exemplary organizational structure distinguishes between customer types

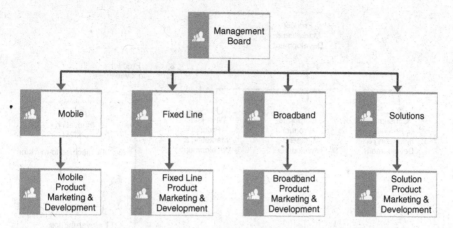

Fig. 4.80 Exemplary organizational structure distinguishes between product types

Another criterion is the product type, which might influence the organizational structure on a high level. Historically grown enterprises sometimes have separate legal entities for mobile and fixed line products. In addition to the network infrastructure, broadband products require additional services (e.g., Internet access) which were originally offered by Internet Service Providers (ISPs). Solutions for communication requirements of special customer groups—for example, corporate customers—are a complex business. Therefore, a possible organizational structure for product marketing and development might distinguish between the different product types on a high level (cf. Fig. 4.80). In this case, each product-based division contains its own product marketing and development unit.

Due to the similarity of organizational criteria of the product and customer-centric domain, an integration of both is possible on a high level. In this case, the organizational entities of sales and customer service as well as product marketing and development could both be combined in an organizational entity, such as a commercial unit (cf. Fig. 4.81).

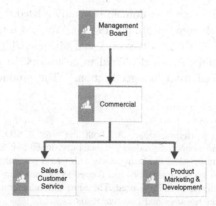

Fig. 4.81 Exemplary organizational structure combines sales and customer service with product marketing and development

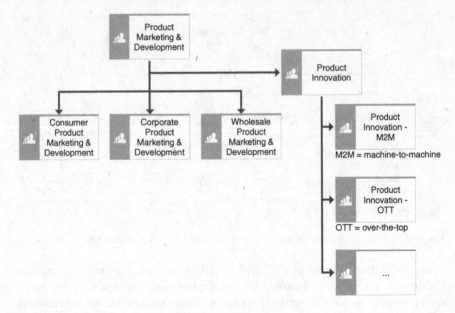

Fig. 4.82 Exemplary organizational structure with central product innovation

In addition to the above criteria, the organizational structure of the product domain might be influenced by innovations such as Machine-to-Machine communication or over-the-top players (cf. Sect. 2.1). In order to react quickly to such new innovations, the creation of organizational entities that are responsible for an overall development of those topics might be advisable. This could be realized by a central product innovation entity (cf. Fig. 4.82).

4.5.3 Data Layer of the Product Domain

The data processed in the product domain is mainly related to the TM Forum SID domain *product* (TM Forum 2015b, pp. 19–22). Products consist of services and resources (Bruce et al. 2008, p. 19; Snoeck and Michiels 2002, p. 335). They are based on market requirements and offered to customers. In addition, a concrete product is distinguished from its specification.[19] The product specification is a

---

[19]The TM Forum uses the term *product* for both the overall SID domain containing all product-related entities and a concrete product instance provisioned to a customer. In addition, the general language might understand the term *product* also as a general term. In order to avoid misunderstandings, in this book the term *product* is used as general term. For a concrete instance of a product the term *concrete product* is used. The same differentiation is used for *service* and *concrete service* as well as *resource* and *concrete resource*.

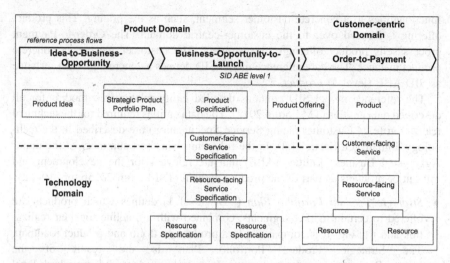

Fig. 4.83 Usage of SID data entities with focus on the product domain

general definition that is used as a blueprint for the provisioning of a concrete product to a customer. The product domain covers the planning, definition, realization, and management of those product specifications. After the realization of a new product, it is handed over to the customer-centric domain. The customer-centric domain is then responsible for selling concrete products to customers. In both cases, the technology domain covers the technical realization through services and resources. This interrelation is illustrated in Fig. 4.83.

The Idea-to-Business-Opportunity process starts with the generation of new product ideas. Based on a first financial and technical evaluation, the further proceeding with a product idea is decided. The product portfolio provides a consolidated view of existing and planned products. In combination with a concrete timeline for realization, a product roadmap is created. Both the product portfolio and product roadmap are included in the SID ABE (level 1) *Strategic Product Portfolio Plan*.

Based on this planning, the Business-Opportunity-to-Launch process designs a product specification. The product specification defines the functionalities of the new product. The technical specification is defined by customer-facing services, resource-facing services, and resources in the technology domain (cf. Sect. 4.4.3). The interface between product domain and technology domain is realized through customer-facing services. Based on the product specification, a product offering is created. The product offering includes further details to offer the defined product functionalities to a concrete customer segment, such as the pricing. This product offering is handed over to the customer-centric domain. The Order-to-Payment process sells product offerings to customers. The customer receives a concrete product instance. The data reference model SID defines a concrete product instance as SID ABE (level 1) *product.*

The reference model SID contains detailed entity relationship models for all described data entities (TM Forum 2012). The data entities required for the technical realization (e.g., Customer-facing Service Specification) are described in the technology domain (cf. Sect. 4.4.3). In the following is a short summary of the major Aggregated Business Entities (ABE) that are relevant for the development and roll-out of products as part of the product domain (TM Forum 2015b, pp. 19–22):

- *Strategic Product Portfolio Plan* (ABE level 1) defines which products are offered to certain market segments, combined with a timeline for their realization and roll-out. It is comparable to a product portfolio and product roadmap. The Strategic Product Portfolio Plan is the result of the Idea-to-Business-Opportunity process. It contains product ideas on a high level that are further detailed in the subsequent processes. The timeline defines the launch of new products, relaunch of existing products, and retirement of existing products.
- *Product Specification* (ABE level 1) contains a functional definition of a product from a business perspective. It is a blueprint for concrete product instances delivered to a customer. The product idea is a result of the Idea-to-Business-Opportunity process and defined in the Strategic Product Portfolio Plan. The Product Specification is created by the Business-Opportunity-to-Launch process. It is mapped to the technical capabilities through Customer-facing Service Specifications (part of the technology domain). A new product specification could be either created through a new arrangement of existing Customer-facing Service Specifications, or it could result in technical requirements for new Customer-facing Service Specifications.
- *Product Offering* (ABE level 1) adds a concrete market offering to a Product Specification (e.g., pricing). It is provided through a product catalog to the customer-centric domain that sells product offerings as concrete products to customers. Product offerings could also define different pricings based on market segments for the same Product Specification.
- *Product* (ABE level 1) contains a concrete instance of a product offering that is sold to a customer. The technical realization of the product is part of the technology domain. As defined in the Product Specification, the product is mapped to a Customer-facing Service.

In addition, the product domain prepares and realizes relaunch and retirement decisions. For these decisions, data about the usage and performance of existing products is required. This is supported by the following SID ABEs (TM Forum 2015b, pp. 19–22):

- *Product Usage* (ABE level 1) provides the usage data for concrete products. It is consolidated based on the Service Usage and Resource Usage provided by the technology domain. It is the primary input for the billing as part of the customer-centric domain. However, it could also provide relevant data to prepare relaunch or retirement decisions.
- *Product Performance* (ABE level 1) includes the performance of (concrete) products. Similar to the Product Usage, it is consolidated based on Service Performance and Resource Performance received from the technology domain. The performance of a product is an important input for relaunch or retirement decisions.

4.5.4 Application Layer of the Product Domain

The processes of the product domain are part of the Business Support Systems (BSS) (Bruce et al. 2008, p. 15; Kelly 2003, p. 109; Snoeck and Michiels 2002, p. 331). A major objective of the BSS is to support the operational pre-sales, sales, and after-sales as part of the customer-centric domain. A prerequisite of these functions is the planning, design, development, and roll-out of products. This is part of the product domain and also covered by the BSS (cf. Fig. 4.84). The application functions of the product domain have an interface to the customer-centric domain. It is included in the BSS and covers the hand-over of new products to the operations. Further interfaces are to the technology domains. These interfaces are between BSS and OSS. They are responsible for the technical realization of products and their operations. In addition, products might involve suppliers. This involvement could be on all levels, i.e., products, services, or resources. In this case, interfaces to the supply chain management systems of the support domain are necessary.

The relevant reference process flows are supported through the TAM domain (level 0) *Product Management* (TM Forum 2015e, p. 58). A more detailed mapping between the reference process flows and TAM application areas (level 1) is provided in Table 4.6. Most of the application areas are used in all four reference process flows, but the focus is different. The Idea-to-Business-Opportunity process requires functions that support the initial part of the product development. At this stage a market requirements, product propositions, and product roadmaps are managed. The required functions are on a strategic and consolidated level. The Business-Opportunity-to-Launch process is responsible for the detailed design as well as the complete realization and roll-out. This process requires more operational functions that allow detailed specifications and modeling as well as project monitoring and management. The two processes Decision-to-Relaunch and Decision-to-Elimination have similar requirements. They are both working on

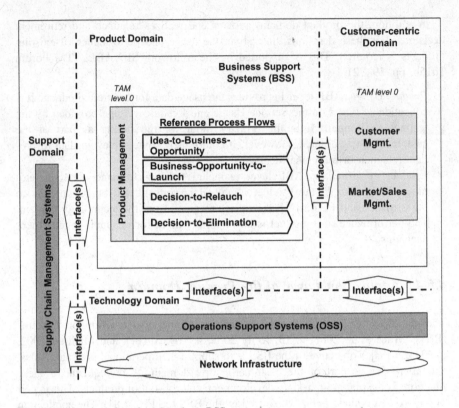

Fig. 4.84 Mapping of product domain to BSS

Table 4.6 Mapping between reference process flows and application areas (level 1) for product domain

Reference process flows	Application areas (Level 1) (TM Forum 2015f, p. 58)[a]
Idea-to-Business-Opportunity	Product strategy/proposition management, product lifecycle management, product performance management
Business-Opportunity-to-Launch	Product catalog management, product lifecycle management, product performance management
Decision-to-Relaunch	Product catalog management, product lifecycle management, product performance management
Decision-to-Elimination	Product catalog management, product lifecycle management, product performance management

[a]The mapping between the reference process flows and application areas is an own proposal. The application areas are based on the high-level illustration of TAM (TM Forum 2015f, p. 58)

existing products. The first step is the preparation of a relaunch or elimination decision which requires performance reports of the existing products. The further steps are the detailed planning, design, and realization of the changes or elimination. The functions for these tasks are similar to the operational functions required in the Business-Opportunity-to-Launch process.

The TM Forum reference model TAM defines these functions in the TAM domain (level 0) *Product Management* as follows (TM Forum 2015e, p. 58):

- *Product Strategy/Proposition Management* (level 1) provides functions for the strategic planning and generation of product ideas. The major functions are the capturing and management of product strategies and propositions combined with reporting and project management tools.
- *Product Catalog Management* (level 1) provides a repository for product specifications. The product catalog is linked to service and resource repositories included in the technology domain (e.g., Service Catalog Management). The realization of an overall catalog management is an important prerequisite for standardized product management processes.
- *Product Lifecycle Management* (level 1) supports the entire product lifecycle from design to retirement. Major functions are the capturing of product requirements, the product specification, the roll-out of a new product, management of existing products, and the retirement of products.
- *Product Performance Management* (level 1) provides functions for the monitoring and reporting of the product development performance as well as the performance of existing products. Those functions cover campaigns, revenue, costs, and capacity.

The product domain combines creativity with management and technical skills. In particular, the idea generation and evaluation part included in the Idea-to-Business-Opportunity process might require further tools and systems. These could include access to external databases and studies, market and technology screenings, and competitor analysis. Furthermore, creativity techniques might be supported by application systems. Based on the innovation level being targeted, the operations of own research laboratories is possible.

4.5.5 Summary of the Product Domain

The above described content of processes, application, data, and organizational mapping is illustrated in a reference architecture for the product domain in Fig. 4.85.

The product domain is responsible for the whole product lifecycle. It starts with the generation and evaluation of product ideas. These product ideas are planned in a product portfolio and product roadmap. Further activities are the specification and roll-out of new products as well as the evaluation and realization of relaunch and retirement of existing products. The focus of the product domain is the management of products from a business perspective. For the technical realization, an interface to the technology domain is required. Products that have been successfully rolled out are handed over to the customer-centric domain.

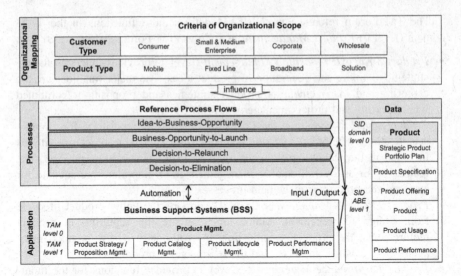

Fig. 4.85 Summary of reference architecture of the product domain

The interrelations to the technology and customer-centric domain are structured through the described concepts of separation between product, service, and resource as well as the differentiation between product specification and concrete product. The technology domain is linked to the product domain through a Customer-facing Service Specification that is mapped to a Product Specification in order to define its technical realization. The customer-centric domain receives a Product Offering as a blueprint for a concrete product instance.

The four reference process flows define the activities required for the whole product lifecycle. The Idea-to-Business-Opportunity process covers the strategic planning of the product portfolio and product roadmap. The detailing and roll-out of new products is managed by the Business-Opportunity-to-Launch process. The relaunch or retirement of existing products is prepared and managed by the Decision-to-Relaunch and Decision-to-Elimination processes. Those processes are supported by the Business Support Systems (BSS). Based on the TM Forum reference model TAM (TM Forum 2015e, p. 58), the relevant functions of the BSS are summarized in the TAM domain product management.

In a concrete implementation, the detailed design of processes, application, and data depends on the organizational scope, which can be summarized based on the dimensions customer type and product type. Both criteria are also relevant for the customer-centric domain. A concrete organizational design requires a close alignment between the product and the customer-centric domain.

4.6   Detailing the Customer Domain and the Support Domain

The customer domain focuses on marketing activities, such as market research or campaigns. In contrast to the customer-centric domain, the customer domain supports customer-related activities—for example, by preparing successful sales through marketing campaigns. The support domain contains all general support activities, including finance or human resource management. For both domains the reference process flows are explained in following sections.

4.6.1   Reference Process Flows of the Customer Domain

The *customer domain* covers those reference process flows[20] that are related to the customer but not initiated by him. It is related to the management of customer relations and sales management (cf. Fig. 4.86).

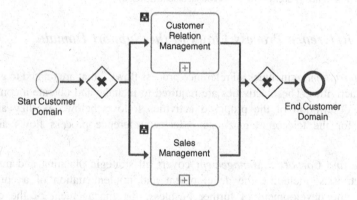

Fig. 4.86   Reference process flows of customer domain (level 2)

Customer Relation Management covers a variety of activities related to the planning, analyzing, handling, and evaluating of customer relations (cf. Fig. 4.87). Some of these activities initiate a direct contact to the customer, such as retention and loyalty. Others support the customer-centric domain—for example, through CRM support and readiness, order handling support, or manage billing events.

---

[20]The reference process flows are based on intensive project work, research and development. This work was mainly conducted by Detecon in various different teams. The two authors had leading roles in this development.

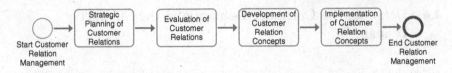

Fig. 4.87 Reference process flow Customer Relation Management (level 2)

Sales Management is related to the sales activities which are operated in the customer-centric domain (cf. Fig. 4.88). It starts with the definition of a sales strategy and ends with the monitoring of sales performance. An important goal is the standardization of sales activities across all platforms, channels, and touch points.

Fig. 4.88 Reference process flow Sales Management (level 2)

4.6.2 Reference Process Flows of the Support Domain

The *support domain* includes all reference process flows[21] that are related to general support activities. These activities are required to manage and operate a company. A short description of the proposed activities follows below. As they are not specific for the telecommunications industry, reference process flows are not provided.

Strategic and Corporate Management covers all strategic planning and management activities, including the development and implementation of a corporate strategy, the development of further business, the management of the overall enterprise architecture, and the standardization across business units.

Financial Management deals with all financial activities from their planning to their operations, including financial planning and budgeting, accounting operations, and asset management.

Human Resource Management deals with all people-related topics to provide the human resources required to operate the enterprise, including organizational development, workforce strategy, and HR policies and practices.

---

[21]The reference process flows are based on intensive project work, research and development. This work was mainly conducted by Detecon in various different teams. The two authors had leading roles in this development.

Supply Chain Management contains all support activities related to the supply chain, including procurement policies, supply chain planning and strategy, logistics support and handling, and procurement operations.

Enterprise Risk Management aims to identify and eliminate possible risks to the enterprise. Those risks are related to different topics, such as security management, fraud management, and business continuity management.

Enterprise Effectiveness Management deals with tools, trainings, and methods related to the assurance and improvement of the enterprise effectiveness, such as process management, quality management, and performance assessment.

Corporate Communications covers the overall management of all internal and external communications.

Legal and Regulatory Management deals with all legal and regulatory requirements.

References

Adamopoulos, D. X., Pavlou, G., & Papandreou, C. A. (2000). Supporting advanced multimedia telecommunications services using the distributed component object model. In J. Delgado, G. D. Stamoulis, A. Mullery, D. Prevedourou, & K. Start (Eds.), *Telecommunications and IT convergence towards service e-volution* (pp. 89–104). Berlin Heidelberg: Springer.

Aier, S., Gleichauf, B., & Winter, R. (2011). Understanding enterprise architecture management design—An empirical analysis. In: A. Bernstein, G. Schwabe (Eds.), *Proceedings of the 10th International Conference on Wirtschaftsinformatik* (pp. 645–654). Zürich.

Aier, S., & Winter, R. (2008). Virtuelle Entkopplung von fachlichen und IT-Strukturen für das IT/business alignment—Grundlagen, Architekturgestaltung und Umsetzung am Beispiel der Domänenbildung. *Wirtschaftsinformatik, 51*, 175–191. doi:10.1007/s11576-008-0115-0

Axenath, B., Kindler, E., & Rubin, V. (2005). An open and formalism independent meta-model for business processes. In: E. Kindler, M. Nüttgens (Eds.), *Proceedings of the Workshop on Business Process Reference Models 2005 (BPRM 2005)* (pp. 45–59). Nancy, France.

Becker, J. (Ed.). (2011). *Process management: A guide for the design of business processes* (2nd ed.). Berlin: Springer.

Becker, J., Berning, W., & Kahn, D. (2005). Projektmanagement. In J. Becker, M. Kugeler, & M. Rosemann (Eds.), *Prozessmanagement* (pp. 17–44). Berlin, Heidelberg: Springer.

Becker, J., & Meise, V. (2005). Strategie und Ordnungsrahmen. In J. Becker, M. Kugeler, & M. Rosemann (Eds.), *Prozessmanagement* (pp. 105–154). Berlin, Heidelberg: Springer.

Bertin, E., & Crespi, N. (2009). Service business processes for the next generation of services: a required step to achieve service convergence. *Annals of Telecommunications, 64*, 187–196.

Betz, C. T. (2011). *Architecture & patterns for IT service management, resource planning, and governance: making shoes for the cobbler's children* (2nd ed.). Waltham, MA: Morgan Kaufmann.

Böhmann, T., Schermann, M., & Krcmar, H. (2007). Application-oriented evaluation of the SDM reference model: Framework, instantiation and initial findings. In J. Becker & P. Delfmann (Eds.), *Reference modeling* (pp. 123–144). Heidelberg: Physica-Verlag.

Brock, G. W. (2002). Historical overview. In: M. E. Cave, S. K. Majumdar, & I. Vogelsang (Eds.), *Handbook of telecommunications economics (volume 1): Structure, regulation and competition* (pp. 43–74). Amsterdam, u. a.: Elsevier.

Bruce, G., Naughton, B., Trew, D., Parsons, M., & Robson, P. (2008). Streamlining the telco production line. *Journal of Telecommunications Management, 1*, 15–32.

Buhl, H. U., & Winter, R. (2008). Full virtualization—BISE's contribution to a vision. *Business and Information Systems Engineering, 1*, 133–136. doi:10.1007/s12599-008-0023-2

Chen, P. P.-S. (1976). The entity-relationship model—Toward a unified view of data, *1*, 9–36. doi:10.1145/320434.320440

Choi, M.-J., & Hong, J. W.-K. (2007). Towards management of next generation networks. IEICE Transactions on Communications, *90–B*, 3004–3014.

Church, J., & Gandal, N. (2005). Platform competition in telecommunications. In S. K. Majumdar, I. Vogelsang, & M. E. Cave (Eds.), *Handbook of telecommunications economics* (Vol. 2, pp. 117–153)., Technology Evolution and the Internet Amsterdam: Elsevier.

Cummings, T. G., & Worley, C. G. (2009). *Organization development & change* (9th ed.). Australia ; Mason, OH: South-Western/Cengage Learning.

Czarnecki, C. (2013). *Entwicklung einer referenzmodellbasierten Unternehmensarchitektur für die Telekommunikationsindustrie*. Berlin: Logos-Verl.

Czarnecki, C., & Spiliopoulou, M. (2012). A holistic framework for the implementation of a next generation network. *International Journal of Business Information Systems, 9*, 385–401.

Czarnecki, C., Winkelmann, A., & Spiliopoulou, M. (2012). Transformation in telecommunication —Analyse und clustering von real-life Projekten. In D. C. Mattfeld & S. Robra-Bissantz (Eds.), *Multi-Konferenz Wirtschaftsinformatik 2012—Tagungsband Der MKWI 2012* (pp. 985–998). Braunschweig: GITO Verlag.

Czarnecki, C., Winkelmann, A., & Spiliopoulou, M. (2013). Reference process flows for telecommunication companies: An extension of the eTOM model. *Business & Information Systems Engineering, 5*, 83–96. doi:10.1007/s12599-013-0250-z

Davies, I., & Reeves, M. (2010). BPM tool selection: The case of the queensland court of justice. In J. vom Brocke & M. Rosemann (Eds.), *Handbook on Business Process Management 1* (pp. 339–360). Berlin Heidelberg: Springer.

Dinsmore, P. C., & Cabanis-Brewin, J. (2014). *The AMA handbook of project management*. New York: AMACOM.

Economides, N. (2005). The economics of the internet backbone. In S. K. Majumdar, I. Vogelsang, & M. E. Cave (Eds.), *Handbook of Telecommunications Economics* (Vol. 2, pp. 373–412)., Technology Evolution and the Internet Amsterdam: Elsevier.

Fleischmann, A. (2013). *S-BPM illustrated: a storybook about business process modeling and execution* (1st ed.). New York: Springer.

Haleplidis, E., Haas, R., Denazis, S., & Koufopavlou, O. (2009). A web service- and ForCES-based programmable router architecture. In D. Hutchison, S. Denazis, L. Lefevre, & G. J. Minden (Eds.), *Active and Programmable Networks* (pp. 108–120). Berlin Heidelberg: Springer.

Hämäläinen, S., Sanneck, H., & Sartori, C. (Eds.). (2012). *LTE self-organising networks (SON): network management automation for operational efficiency*. Hoboken, N.J: Wiley.

Hammer, M., & Champy, J. (1994). *Reengineering the corporation: A manifesto for business revolution*. New York, NY: HarperBusiness.

Hanrahan, H. (2007). *Network convergence services, applications, transport, and operations support*. Chichester, England; Hoboken, NJ: John Wiley & Sons.

Hill, G. (2007). *The cable and telecommunications professionals' reference*. Amsterdam, Boston: Focal Press.

Hunter, M. G. (2011). The duality of information technology roles: A case study. *International Journal of Strategic Information Technology and Applications, 2*, 37–47. doi:10.4018/jsita.2011010103

Iannone, E. (2012). *Telecommunication networks, devices, circuits, and systems*. Boca Raton, FL: CRC Press.

ITU. (2007a). ITU-T Recommendation M.3050.1: Enhanced Telecom Operations Map (eTOM)—The business process framework.

ITU. (2007b). ITU-T Recommendation M.3050.2: Enhanced Telecom Operations Map (eTOM)—Process decompositions and descriptions.

ITU (Ed.). (2008a). ITU-T Recommendation M.3190: Shared information and data model (SID).

ITU. (2008b). ITU-T Recommendation L.80: Operations support system requirements for infrastructure and network elements management using ID technology.

Jaya Shankar, P., Chandrasekaran, V., & Desikan, N. (2000). An agent based service inter-working architecture for the virtual home environment. In J. Delgado, G. D. Stamoulis, A. Mullery, D. Prevedourou, & K. Start (Eds.), *Telecommunications and IT Convergence Towards Service E-Volution* (pp. 257–268). Berlin Heidelberg: Springer.

Jones, G. R. (2013). *Organizational theory, design, and change* (7th ed.). Upper Saddle River, NJ: Pearson.

Kelly, M. B. (2003). The telemanagement forum's enhanced telecom operations map (eTOM). *Journal of Network and Systems Management, 11*, 109–119.

Knightson, K., Morita, N., & Towle, T. (2005). NGN architecture: Generic principles, functional architecture, and implementation. *IEEE Communications Magazine, 43*, 49–56. doi:10.1109/MCOM.2005.1522124

Kusnetzky, D. (2011). *Virtualization: A manager's guide*. Sebastopol, CA: O'Reilly.

Laudon, K. C., & Laudon, J. P. (2012). *Management information systems: managing the digital firm* (12th ed.). Boston: Prentice Hall.

Lewis, L. (2001). *Managing business and service networks*. New York, u. a.: Kluwer Academic Publishers.

Long, J. (2014). *Process modeling style*. Waltham, MA: Kaufmann.

Mansfield, R. (2013). *Company strategy and organizational design*. Hoboken: Taylor and Francis.

Matthes, D. (2011). *Enterprise architecture frameworks kompendium*. Heidelberg, u. a.: Springer.

McAuley, J., Duberley, J., & Johnson, P. (2007). *Organization theory: Challenges and perspectives*. Harlow, England, New York: Prentice Hall/Financial Times.

Mikkonen, K., Hallikas, J., & Pynnönen, M. (2008). Connecting customer requirements into the multi-play business model. *Journal of Telecommunications Management, 2*, 177–188.

Mishra, A. R. (2007). *Advanced cellular network planning and optimisation 2G/2.5G/3G—evolution to 4G*. Chichester: John Wiley.

Misra, K. (2004). *OSS for telecom networks: An introduction to network management*. London, u. a.: Springer.

OMG. (2011). Business process model and notation (BPMN)—Version 2.0.

Orand, B. (2013). Foundations of IT service management: With ITIL 2011.

Porter, M. E. (2004). *Competitive advantage*. New York, London: Free.

Rockart, J. F., Earl, M. J., & Ross, J. W. (2003). Eight imperatives for the new IT organization. In T. W. Malone, R. Laubacher, & M. S. S. Morton (Eds.), *Inventing the Organizations of the 21st Century*. : MIT Press.

Rosemann, M. (2003). Preparation of process modeling. In J. Becker, M. Kugeler, & M. Rosemann (Eds.), *Process Management* (pp. 41–78). Berlin Heidelberg, Berlin, Heidelberg: Springer.

Scheer, A.-W. (1997). *Wirtschaftsinformatik: Referenzmodelle für industrielle Geschäftsprozesse*. Berlin, u. a.: Springer.

Scheer, A.-W. (2000). *ARIS–business process modeling* (3rd ed.). Berlin ; New York: Springer.

Scheer, A.-W., Jost, W., & Öner, G. (2007). A reference model for industrial enterprises. In P. Fettke & P. Loos (Eds.), *Reference Modeling for Business Systems Analysis* (pp. 167–181). Hershey, PA: Idea Group Publishing.

Schekkerman, J. (2004). *How to survive in the jungle of enterprise architecture frameworks: creating or choosing an enterprise architecture framework*. Victoria: Trafford.

Schütte, R. (1998). *Grundsätze ordnungsmässiger Referenzmodellierung: Konstruktion konfigurations- und anpassungsorientierter Modelle*. Wiesbaden: Gabler.

Sharkey, W. W. (2002). Representation of technology and production. In M. E. Cave, S. K. Majumdar, & I. Vogelsang (Eds.), *Handbook of Telecommunications Economics (Volume 1): Structure, Regulation and Competition* (pp. 179–222). Amsterdam, u. a.: Elsevier.

Snoeck, M., & Michiels, C. (2002). Domain modelling and the co-design of business rules in the telecommunication business area. *Information Systems Frontiers, 4,* 331–342.

Stair, R. M., & Reynolds, G. W. (2012). *Fundamentals of information systems.* Boston: Course Technology/Cengage Learning.

The Open Group. (2011). TOGAF Version 9.1. Zaltbommel: Van Haren Publishing.

TM Forum (Ed.). (2010). eTOM (Business Process Framework) GB921 Addendum E: End-to-End Business Flows (Version 9.0).

TM Forum (Ed.). (2012). Information Framework (SID) GB922 0-P: Information Framework Primer (Version 7.3).

TM Forum. (2015a). Business process framwork (eTOM): Concepts and principles (GB921 CP), Version 15.0.0. ed.

TM Forum. (2015b). Information framwork (SID): Concepts and principles (GB922), Version 15.0.0. ed.

TM Forum. (2015c). Application framwork (TAM): Concepts and principles (GB929 CP), Version 14.5.1. ed.

TM Forum. (2015d). Business process framwork (eTOM): End-to-end business flows (GB921 Addendum E), Version 15.0.0. ed.

TM Forum. (2015e). Application framwork: The digital services systems landscape (GB929 Addendum D), Version 14.5.1. ed.

TM Forum. (2015f). GB921 process framework primer (Version 15).

Tsukas, H., & Knudsen, C. (Eds.). (2005). *The Oxford handbook of organization theory: [meta-theoretical perspectives].* Oxford: Oxford University Press.

Urbaczewski, L., & Mrdalj, S. (2007). A comparison of enterprise architecture frameworks. *Issues in Information Systems, 7,* 18–23.

Ward, J., & Peppard, J. (2002). Strategic planning for information systems. *Wiley series in information systems* (3rd ed.). Chichester, West Sussex, England ; New York: J. Wiley.

Wellenius, B., & Townsend, D. N. (2005). Telecommunications and economic development. In S. K. Majumdar, I. Vogelsang, & M. E. Cave (Eds.), *Handbook of Telecommunications Economics* (Vol. 2, pp. 555–619)., Technology Evolution and the Internet Amsterdam: Elsevier.

Westland, J. (2007). *The project management life cycle: A complete step-by-step methodology for initiating, planning, executing & closing a project successfully,* Repr. ed. London [u.a.]: Kogan Page.

Wigand, R. T., Mertens, P., Bodendorf, F., König, W., Picot, A., & Schumann, M. (2003). *Introduction to business information systems.* Berlin, New York: Springer.

Winter, R., & Fischer, R. (2007). Essential layers, artifacts, and dependencies of enterprise architecture. *Journal for Enterprise Architecture, 2,* 7–18.

Yahia, I. G. B., Bertin, E., & Crespi, N. (2006). Next/new generation networks services and management. In *Proceedings of the International Conference on Networking and Services, ICNS'06* (p. 15). Washington, DC, USA: IEEE Computer Society. doi:10.1109/ICNS.2006.77

Zuboff, S. (1988). *In the age of the smart machine: the future of work and power.* New York: Basic Books.

Chapter 5
Planning and Implementing the Architecture Solution

Abstract Planning and implementing the architecture solution is essential to benefit from the solution design. From a dynamic perspective the architectural implementation is a transformation from the current state of the enterprise to a targeted state that is defined by the solution design. In most cases, the entire design and implementation are conducted in a cross-functional project. With respect to the duration and persons involved, such a project can be seen as complex endeavor. Various interrelations between the architectural elements, conflicts of objective between different organizational entities, and changing external or internal factors require careful consideration. For planning the tasks from the set-up to design and implementation, an Architecture Solution Map is proposed (cf. Sect. 5.1). It consists of eight major tasks: architecture diagnostics, strategic alignment, architecture framework, architecture ownership, architecture design, training and awareness, change management, and architecture implementation. Detailed recommendations and guidelines based on numerous experiences with real-life transformation projects in the telecommunications industry are discussed. Furthermore, transformation types and organizational responsibilities (cf. Sect. 5.2), typical project examples (cf. Sect. 5.3), and detailed case studies (cf. Sect. 5.4) are provided.

Cost reduction, revenue increase, and quality improvement are usually the strategic objectives and key drivers for planning and implementing an architecture solution. In fact, the transformation of a telecommunications operator has to be carefully planned in order to achieve the desired results. The detailed architecture solution described in Chap. 4 can be used as a reference model for designing an architecture solution of a telecommunications operator. It provides detailed reference solutions from a topical perspective. A simplified version of this detailed architecture solution is illustrated in Fig. 5.1 and contains the relevant architecture layers to be considered.

From a methodical perspective the planning, designing, and implementing of an architecture solution or the changing of an existing architecture solution of a telecommunications operator should be performed by applying a standardized approach. Such a transformation endeavor is typically a complex task. In this

© Springer International Publishing AG 2017
C. Czarnecki and C. Dietze, *Reference Architecture for the Telecommunications Industry*, Progress in IS, DOI 10.1007/978-3-319-46757-3_5

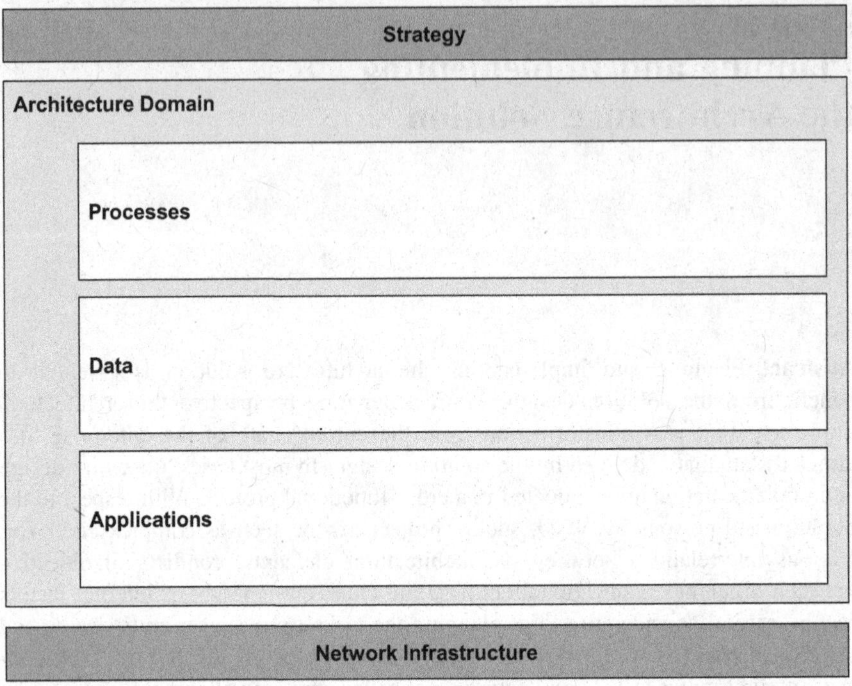

Fig. 5.1 Simplified architecture solution

context, various general guidance is provided by publications about project and program management (e.g. Brown 2008; Dinsmore and Cabanis-Brewin 2014; Harvard Business Review Press 2013; Westland 2007), transformation and change management (e.g. George 2006; Koenigsaecker 2013; Kotter 2007), organizational development (e.g. Cummings and Worley 2009; Jones 2013), and information systems development (e.g. Laudon and Laudon 2012; Ward and Peppard 2002; Wigand et al. 2003). Furthermore, existing methods of Enterprise Architecture Management (EAM) can be utilized (e.g. Ahlemann 2012; Lankhorst 2013; Van Den Berg and Van Steenbergen 2006). Due to the complexity of this topic, the major challenge in practice is a manageable starting point for structuring the required work. Therefore, the purpose of this chapter is not to provide a new method for transformation, project, or program management, but to structure and discuss major lessons learned from real-life projects with a focus on the telecommunications industry.

In Sect. 5.1, an *Architecture Solution Map* is introduced that structures the major tasks of an architectural transformation. The planning, designing, and implementing of an architecture solution are usually performed as part of a transformation program, and the *Architecture Solution Map* therefore also reflects the important transformation program management aspect. Three general architecture solution transformation types (i.e., strategic, technical and operational) related to the telecommunications industry together with concrete project examples for each transformation type are described in Sect. 5.2. In this respect, the importance for the involvement of top management representatives is highlighted, and a guideline for top management organizational responsibilities related to each transformation type is provided. To underline the practical relevance of this book, ten concrete examples of different transformation projects are provided in Sect. 5.3, out of which four concrete projects are described as detailed example cases in Sect. 5.4.

Most of the concepts introduced in Chapter 5 are derived from the results of projects that were performed by Detecon International GmbH[1] in the international telecommunications industry. The two authors were personally involved in these projects and have a detailed knowledge about the scope of work and results of these projects.

5.1 Architecture Solution Map

For planning, designing, and implementing the architecture solution introduced in Chap. 4, a standardized, step-wise approach is recommended. In this section, the focus is on the elaboration of an *Architecture Solution Map* that can be used by telecommunications operators as a reference for structuring the tasks related to architecture solution planning, designing, and implementing. This *Architecture Solution Map*[2] has been developed with a focus on the telecommunications industry, and it has successfully been applied in project engagements with international telecommunications operators.

The *Architecture Solution Map* consists of eight key elements and one overarching *Transformation Program Management* element. An overview of the eight key elements and the overarching *Transformation Program Management* element is shown in Fig. 5.2. The selection and also the application order of the elements can be customized according to the specific situation. A telecommunications operator that applies the *Architecture Solution Map* should determine the most suitable sequence of the elements according to their specific project circumstances; for example, an operational project for the introduction of an Operational Support System (OSS) might not require a strategic alignment.

The *Architecture Solution Map* as depicted in Fig. 5.2 can be applied as part of a transformation program and therefore a separate overarching *Transformation Program Management* element is introduced. Comprehensive international program management standards are provided, e.g., by the Program Management

---

[1]cf. www.detecon.com

[2]The fundamental idea of the Architecture Solution Map elaborated in this section is derived from an existing Business Process Management Map that has been developed by the two authors in the context of international project work with Detecon.

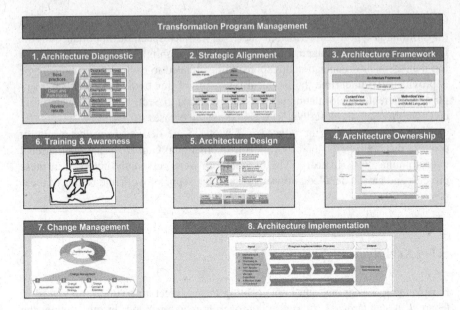

Fig. 5.2 Architecture solution map elements

Institute[3] (PMI) and PRINCE2.[4] Furthermore, various general publications exist (e.g. Brown 2008; Dinsmore and Cabanis-Brewin 2014; Harvard Business Review Press 2013; Westland 2007). It is recommended to consider one of those comprehensive projects and program management concepts for a transformation program. As first indication for the complexity of the required work, some examples for breaking down a transformation program and for developing a transformation program plan are discussed in this section.

Executing a transformation requires a detailed breakdown of the whole transformation program. In particular, the complexity of a transformation presents a typical challenge. Therefore, a breakdown of the whole transformation program into manageable parts is recommended. A general, hierarchical structure on how to break down a complex transformation program into activities is illustrated in Fig. 5.3:

- Step 1: The *transformation program* is structured into *transformation projects*.
- Step 2: Each *transformation project* is structured into *sub-projects*.
- Step 3: Each *sub-project* is structured into *work streams*.
- Step 4: Each *work stream* is structured into *work packages*.
- Step 5: Each *work package* is structured into detailed *activities*.

---

[3]cf. www.pmi.org

[4]cf. www.prince2.com

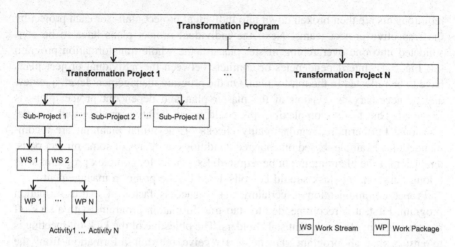

Fig. 5.3 Transformation program breakdown

This hierarchical structure of the transformation program should then be related to responsibilities which reflect the program organization. On a high level, the program manager is supported by project managers that are responsible for the different transformation projects. Further responsibilities are defined on the operational level with project teams (cf. Sect. 5.1.4).

Furthermore, the breakdown of the transformation program into transformation projects, sub-projects, work streams, work packages, and activities is used for developing a detailed project plan. The work breakdown structure combined with the different responsibilities should be reflected in an overall project plan. In addition, the project plan includes the estimated duration of each project phase, work stream, work package and activity. Project milestones indicate the point of time where important activities have to be concluded and pre-defined project deliverables have to be provided. Project deliverables have to be defined as part of the project planning and have to be linked to project phases, work streams, work packages or activities.

Typically, the monitoring of the program progress is conducted by different committees. The high-level milestone and major deliverables are controlled by the *Program Steering Committee* that consists of top-management representatives as well as the project managers of the different transformation projects. Also, on the sub-project, work stream, and work packaged levels, regular progress meetings and reports are required. In this context, a consistent management and communication from the high-level program perspective to the operational work packages is a challenge that is typically supported by a program management office.

Transparency of the different program parts and their interrelation can be achieved by the development of a master plan. A combination of both a top-down and bottom-up approach is recommended. First general requirements (e.g., high-level milestones) should be defined on the program management level. Those requirements are then broken down into individual project plans for each project by the respective project teams. Next, the individual project plans have to be consolidated into one overarching master plan for the whole transformation program (cf. Fig. 5.4). Interdependencies or conflicts between the individual project plans have to be reflected in the master plan. On the program management level, a review and, if necessary, a revision of the master plan and dependent project plans is triggered. Due to the complexity, this could be an iterative process with several revisions. Furthermore, regular reality checks of the initial planning are recommended: for example, based on changed conditions or delays of some project parts, a revision of the planning might be required. Especially for complex programs with a long duration, this task should be considered by the program management.

Proper communication is certainly a key success factor of a transformation program. First, it is recommended to start transformation programs with a kick-off meeting that involves relevant stakeholders. The objective of the kick-off meeting is to ensure that all program stakeholders receive relevant information from the program manager who has the responsibility and the program sponsor who has the accountability for the transformation program. During the kick-off meeting, all of the program objectives, the work breakdown structure, roles and responsibilities, the program plan, the expected deliverables, potential risks, and communication as

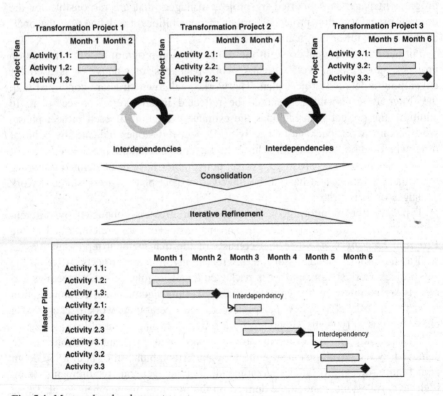

Fig. 5.4 Master plan development

well as escalation rules are presented to the audience. Due to the complexity of a transformation program, the preparation of that information might require substantial preliminary work. This could be part of preliminary project that already involves major stakeholders (e.g., project managers).

Operational project management tools like reporting, activity lists, risk and issue registers, meeting schedules, and status calls also have to be implemented as part of the program management. The early involvement of functional owners in workshops is critical for the development of the pre-defined transformation program deliverables. Through the introduction of control gates after the finalization of each project phase, the work stream and work package, as well as the quantity and quality of the project deliverables can all be assessed. If the quantity or quality of the project deliverables is not compliant with the pre-defined target and standards, specific optimization measures have to be defined and implemented.

Each of the eight key elements of the *Architecture Solution Map* has a concrete scope of work and a set of standard tools to be applied. Further details for each of the eight map elements are provided in the following part of this section.

5.1.1   Map Element 1: Architecture Diagnostics

The objective of the Architecture Diagnostics element is the identification of existing gaps in the current enterprise situation (as-is). In this respect, interviews and workshops with experts of the different architecture layers (i.e., strategy, processes, data, applications and network infrastructure) are important in order to get transparency about the strengths and weaknesses. In addition, available documents (e.g., process charts, data models) providing further details should be thoroughly reviewed. In this context, it is important to differentiate between the implemented as-is situation and the existing documents. Gaps between those two perspectives are an important finding for the diagnostic study; as an example, it is possible that the as-is situation differs from the documented situation or that some parts are not documented at all.

Furthermore, it is essential to determine the maturity level of each architecture layer, but also if there is sufficient alignment and linkage between the different layers. A comparison of the as-is situation with existing reference solutions (e.g. Chap. 4) and industry standards (e.g. TOGAF, eTOM, ITIL) should also be considered. Those reference solutions are also helpful to identify the relevant experts of the enterprise. In this context, the high-level reference architecture (cf. Sect. 4.1) is a good starting point. In a first step, the relevant organizational entities are mapped to the reference architecture, and experts for interviews and workshops are identified. In a second step, gaps between the reference architecture and the as-is situation could be identified, which might be additional findings of the diagnostic study.

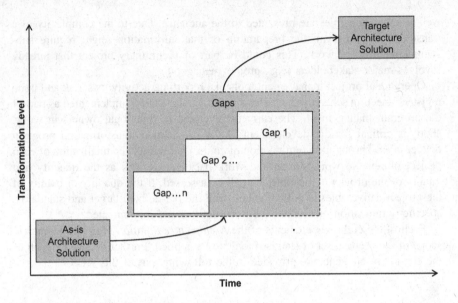

Fig. 5.5 Architecture gap analysis

The expected outcome of the *Architecture Diagnostic* element is a list of existing gaps related to the current situation compared to the targeted architecture solution (cf. Fig. 5.5). Generally, this leads to the challenge that the detailed target architecture is not known at this stage as it requires a detailed design (cf. Sect. 5.1.5). However, the identification of gaps is essential for estimating the effort and planning the further activities. A typical solution is the usage of existing reference solutions and industry standards. Furthermore, target setting workshops are recommended to identify specific requirements. The combination of all gaps basically determines the gap between the as-is situation and the desired target architecture solution of the telecommunications operator. The detailed design and implementation of this target architecture is then the objective of the whole transformation program.

For each gap of the current situation, a proper description and impact estimation for gap closure should be provided (cf. Fig. 5.6).

Once the existing gaps between the as-is situation and the target architecture solution are identified, it is important to perform a prioritization. It is not realistic to start with the closure of all existing gaps at the same time, but rather to focus on the realization of quick wins as part of the *Architecture Diagnostic* element. The prioritization of gaps and respectively the identification of quick wins can be achieved using a matrix with the two dimensions *effort* and *impact* as depicted in Fig. 5.7. On the one hand, those gaps that can be closed with a low realization effort and have a high impact are prioritized the highest and can be characterized as quick wins. On the other hand, those gaps that can only be closed with a significantly higher effort and have an expected low impact should be given little priority.

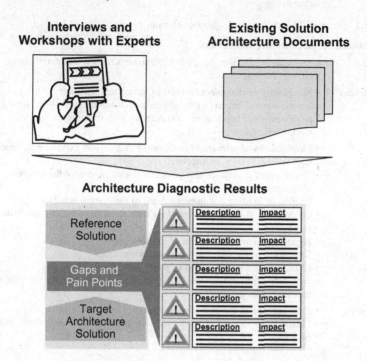

Fig. 5.6 Architecture diagnostic element

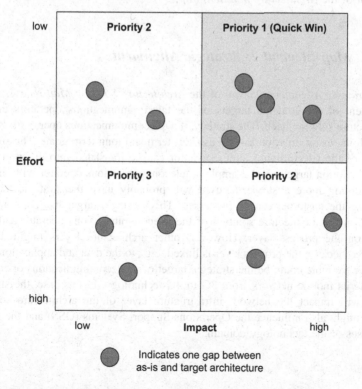

Fig. 5.7 Gap prioritization matrix

Table 5.1 Summary of architecture diagnostics element

Map Element 1: Architecture Diagnostics	
Objective	Identification of existing gaps in the current situation and definition of measures for gap closure
Prerequisites	• Availability of documents related to the current situation • Involvement of experts of the different architecture layers (i.e., strategy, processes, data, applications and network infrastructure) in interviews and workshops
Main activities	• Identification of relevant organizational entities and experts, e.g., based on high-level reference architecture • Execution of interviews and workshops with experts of the different architecture layers • Review of existing documents related to the current situation • Comparison of the current situation with reference solutions and industry standards • Identification and description of existing gaps • Estimation of the effort and impact for each gap • Evaluation and prioritization of gaps • Summary of diagnostic results including a recommendation for required transformation measures
Key results	• Transparency of the current situation • Detailed overview of existing gaps with an evaluation and prioritization of each gap

Table 5.1 summarizes the most important aspects of the *Architecture Diagnostic* element of the *Architecture Solution Map*.

5.1.2   Map Element 2: Strategic Alignment

The *Strategic Alignment* element of the *Architecture Solution Map* focuses on the alignment of the strategic targets of the telecommunications operators and the architecture solution itself. The strategy of a telecommunications operator is defined through its vision, mission as well as short term and long term goals. The strategic targets of the telecommunications operator can be translated into different architecture solution targets. For example, a telecommunications operator with the goal of becoming more customer-focused will probably have the strategic target to optimize the customer-centric processes. This strategic target has an immediate impact on the architecture solution of the telecommunications operator and especially on the process layer. However, other architecture layers might also be impacted because the process layer is linked, e.g., to the data and application layer. Another example could be the strategic target of a telecommunications operator to upgrade its mobile network from 3G to 4G technology. In this case, the strategic target will impact the network infrastructure layer of the architecture solution, which might also influence the Operations Support Systems (OSS) and the related processes of the technology domain.

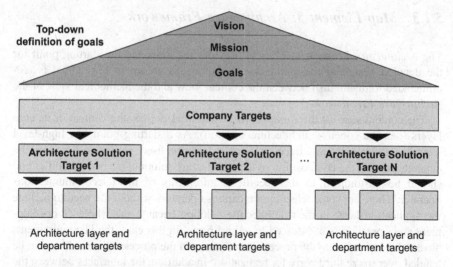

Fig. 5.8 Alignment of strategic targets and architecture solution

Each architecture target that is derived from the strategic targets of the telecommunications operator (cf. Fig. 5.8) is then translated into operational targets for the architecture layers processes, data, applications and network infrastructure. The targets for the different architecture layers should be planned and realized through the organizational entities that are mainly in charge of these layers; for example, the IT department might be responsible for realizing targets that are related to the applications layer, whereas the technology department might be responsible for realizing targets that are related to the network infrastructure layer.

Table 5.2 summarizes the most important aspects of the *Strategic Alignment* element of the *Architecture Solution Map*.

Table 5.2 Summary of strategic alignment element

Map Element 2: Strategic Alignment	
Objective	Alignment of the strategic targets of the telecommunications operator and the architecture solution
Prerequisites	• Transparency of the strategic targets of the telecommunications operator • Consistent mission and vision of the telecommunications operator
Main activities	• Translation of the strategic targets of the telecommunications operator into different architecture solution targets • Translation of architecture solution targets into specific targets for the architecture layers processes, data, applications and network infrastructure • Planning and realization of the different architecture layer targets through the departments in charge of the respective layers
Key results	• Strategic targets of the telecommunications operator aligned with the architecture solution • Architecture solution targets translated into specific targets for the different architecture layers

5.1.3   Map Element 3: Architecture Framework

The *Architecture Framework* element provides the basis and the starting point for the detailed design of the architecture solution. As shown in Fig. 5.9, it is recommended to distinguish between the content view and the methodical view of the *Architecture Framework*.

The content view of the *Architecture Framework* defines the content of its core layers (i.e., the concrete architecture solution). As a starting point, the high-level reference architecture introduced in Sect. 4.1 can be used. However, based on the concrete context the composition of the architecture solution layers might differ and should be customized to the specific requirements of the telecommunications operator. There are some telecommunications operators without the need to include the application layer in their architecture solution because the whole IT development and operation is outsourced to a third party. In this case, the IT requirements have to be derived from the process descriptions in the process layer and have to be handed over to the third party for realization. In addition, the interfaces between the processes, the data and the network infrastructure have to be aligned with the IT outsourcing partner. Hence, in this example there is no need for designing a detailed application layer as part of the architecture solution.

An important objective of the architecture framework is a clear and simple summary of the major architecture parts. It should be self-explanatory, as it is used as a communication instrument for the different stakeholders of the transformation. It should help to understand the overall scope and possible interdependencies. Furthermore, it serves as a high-level target picture. Therefore, it is recommended to carefully design and illustrate the architecture framework. Based on the complexity of the whole architecture solution, a further breakdown into different frameworks per layer might be necessary. In Fig. 5.10, an exemplary framework of the process layer is shown that is based on the reference process flows described in Sect. 4.1.3.

The methodical view of the *Architecture Framework* (cf. Fig. 5.9) defines the documentation standards, the model languages and the design tools to be used for each layer of the architecture solution. With respect to the example of the process layer, the documentation standard and the modeling language (i.e., the process flow notation) provide guidelines for the number of process levels to be designed, the visualization symbols for processes, the design method for interfaces, the allocation

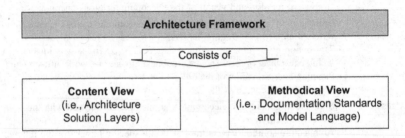

Fig. 5.9 Architecture framework views

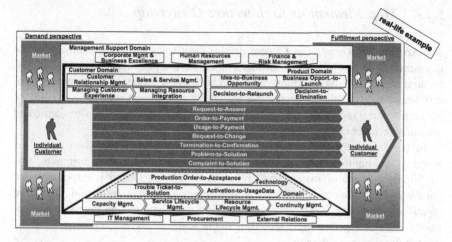

Fig. 5.10 Exemplary framework of the process layer

of Key Performance Indicators (KPIs) to process steps, etc. Amongst the most common process flow notations are the Business Process Model and Notation (BPMN), the Event-Driven Process Chains (EPC), and the subject-oriented Business Process Management (S-BPM) as described in Sect. 4.2.1.

In this section, only examples for the process layer as part of the *Architecture Framework* are provided. For the completion of the *Architecture Framework,* detailed concepts related to the content view and the methodical view for all layers of the architecture solution are provided in Chap. 4.

Table 5.3 summarizes the most important aspects of the *Architecture Framework* element of the *Architecture Solution Map.*

Table 5.3 Summary of architecture framework element

Map Element 3: Architecture Framework	
Objective	Development of an architecture framework providing the starting point for the detailed design of the architecture solution
Prerequisites	• Strategic targets of the telecommunications operator aligned with the architecture solution • Architecture solution targets translated into specific targets for the different architecture layers
Main activities	• Definition of the core layers of the architecture solution (e.g., strategy, processes, data, applications and network infrastructure) as part of the content view • Definition of specific parts for each core layer of the architecture solution (which might be divided into different frameworks for each layer) • Definition of documentation standards, model languages and selection of design tools for each layer of the architecture solution as part of the methodical view
Key results	• Core layers of the architecture solution defined • Specific architecture parts for each core layer elaborated • Documentation standards, model languages and design tools selected for each core layer

5.1.4 Map Element 4: Architecture Ownership

The *Architecture Ownership* element defines clear roles and responsibilities in the organization of the telecommunications operator for planning, designing, implementing and continuously improving the architecture solution. It is recommended to distinguish between the ownership for the whole architecture solution consisting of all architecture layers on the one hand, and the ownership for one dedicated architecture layer on the other hand (cf. Fig. 5.11).

In the first step, the responsible top management representatives (e.g., the Steering Committee of the transformation program) should nominate one *Architecture Solution Owner*. The *Architecture Solution Owner* is responsible for the methodical governance of the whole architectural transformation, including identifying, defining and maintaining the high-level architecture layers (e.g., processes, data, and applications) according to the actual needs and requirements of the telecommunications operator.

For each layer of the defined *architecture solution*, one dedicated *Architecture Layer Owner* should be nominated by the responsible top management representatives. *Architecture Layer Owners* are accountable for the overall management of the detailed design, implementation and continuous improvement of their respective architecture layer. In particular, the alignment of the architecture solution across their respective architecture layer is their responsibility. The owner of the process layer has the overall accountability for the design of detailed processes down to an operational level, the implementation of processes through trainings, as well as addressing required IT system changes, the performance management of processes, and the continuous improvement of processes.

Architecture Layer Owners who are accountable for their respective architecture layer are supported by functional experts in the organization of the telecommunications operators who are actually responsible for executing the required tasks on an operational level. The Architecture Layer Owner for the process layer might be supported by process owners and sub-process partners (cf. Sect. 4.2.2). This detailed definition of architecture ownership links the architecture elements and the organizational responsibilities. Depending on the concrete organizational structures, this task requires a specific definition. The organizational mapping described as part

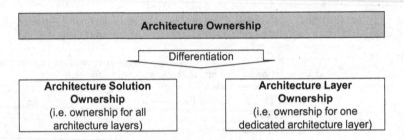

Fig. 5.11 Architecture solution versus architecture layer ownership

of the reference architecture solution provides a starting point for this task. From a methodical perspective, Sect. 4.2.2 provides detailed recommendation for structuring the organizational mapping. References for a mapping between the architecture domains and the organizational structure are described from a process perspective in the respective sections of the reference architecture (cf. Sects. 4.3.2, 4.4.2 and 4.5.2).

In Fig. 5.12 the key tasks of the *Architecture Solution Owner* and the *Architecture Layer Owners* are summarized. It is important that the *Architecture Layer Owners* have to be supported by additional functional experts in the organization, and these have to be identified through an organizational mapping. Sections 4.3.2, 4.4.2 and 4.5.2 provide organizational mapping examples that are driven by the customer centric domain, the technology domain and the product domain of the reference architecture.

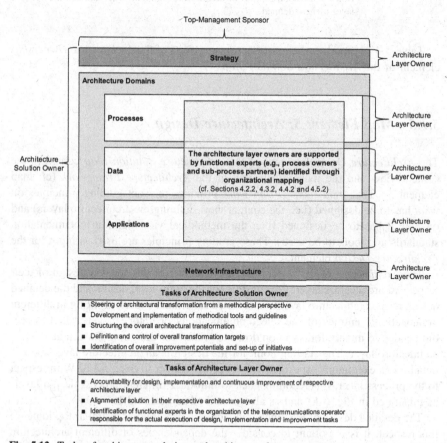

Fig. 5.12 Tasks of architecture solution and architecture layer owner

Table 5.4 Summary of architecture ownership element

Map Element 4: Architecture Ownership	
Objective	Define clear roles and responsibilities in the organization of the telecommunications operator for planning, designing, implementing and continuously improving the architecture solution
Prerequisites	• Core layers of the architecture solution defined • Specific architecture parts for each core layer defined
Main activities	• Distinction between architecture solution ownership and architecture layer ownership • Nomination of one architecture solution owner by the of the telecommunications operator • Nomination of dedicated architecture layer owners by the top management • Definition of clear roles and responsibilities of the architecture solution owner and the architecture layer owners
Key results	• Architecture solution owner and architecture layer owners nominated • Roles and responsibilities of architecture solution owner and architecture layer owners defined

Table 5.4 summarizes the most important aspects of the *Architecture Ownership* element of the *Architecture Solution Map*.

5.1.5   Map Element 5: Architecture Design

The *Architecture Design* element of the *Architecture Solution Map* considers the content view and the methodical view of the *Architecture Framework* (cf. map element 3). The *Architecture Framework* provides relevant guiding principles on *what* has to be designed (i.e., the content view defining the architecture layers) and on *how* it has to be designed (i.e., the methodical view defining documentation standards and tools to be used). These guiding principles are used as input for the *Architecture Design* element.

The *Architecture Layer Owners* are accountable for the detailed design of their respective architecture layer. Responsible for the high- level design and the detailed design of the architecture layers are selected functional experts working in different organizational entities of the telecommunications operator. In general, it is recommended to mainly follow a top down approach for designing the details of each architecture layer. The starting point for the design is an architecture layer-specific detailing as exemplarily specified for the process layer in Sect. 5.1.3. With respect to the process layer, it is recommended to introduce different process design levels as illustrated in Fig. 5.13 and as also described in Sect. 4.2.1.

The detailed design of the other architecture layers follows a similar top logic. In this respect, it is important to consider the dependencies of different architecture layers as described in Chap. 4 for the customer-centric domain (cf. Sect. 4.3), technology domain (cf. Sect. 4.4), product domain (cf. Sect. 4.5), and customer and support domains (cf. Sect. 4.6).

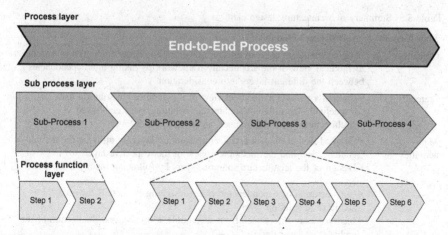

Fig. 5.13 Architecture layer design example (illustrative)

While designing the details of an architecture layer, several iterations might be required to finalize the design. In Fig. 5.14, the high-level, structured approach for the development of the draft architecture layer design, the incorporation of feedback, the finalization, and the approval of the architecture layer design is summarized. In this context, the prior definition of the relevant stakeholders, as provided by the architecture ownership (cf. map element 4), is an important prerequisite. Furthermore, the exact rules for feedback and decisions should be clearly defined and communicated beforehand.

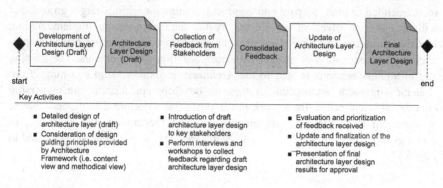

Fig. 5.14 Architecture layer design approval process

Table 5.5 Summary of architecture design element

Map Element 5: Architecture Design	
Objective	Detailed design of each architecture layer based on the content view and the methodical view of the architecture framework by taking interdependencies between the different layers into consideration
Prerequisites	• Content view and methodical view of the architecture solution and respectively the different architecture solution layers elaborated • Architecture ownership defined on an operational level
Main activities	• Detailed design of each architecture solution layer by utilizing the defined content view and the methodical view of the respective layer • Design of the architecture solution layers by following mainly a top-down approach • Consideration of dependencies while designing the details for each architecture solution layer • Incorporation of feedback from stakeholders, update, and finalization of the architecture layer design
Key results	• Detailed design of architecture solution layers compliant with content and methodical view • Feedback from stakeholders collected and incorporated in final design results

Table 5.5 summarizes the most important aspects of the *Architecture Design* element of the *Architecture Solution Map*.

5.1.6 Map Element 6: Training and Awareness

While the detailed design of the different architecture layers is in progress, it is recommended to plan, prepare and conduct trainings for various target groups that will be either involved in or affected by the later architecture implementation. However, training and awareness sessions should optimally be performed in parallel with the execution of most *Architecture Solution Map* elements. The training and awareness sessions related to the *Architecture Solution Map* can range from strategic alignment, architecture framework development, architecture ownership conception, architecture framework development, architecture design, and change management to architecture implementation. The preparation and execution of trainings requires the consideration of several elements as exemplarily illustrated in Fig. 5.15.

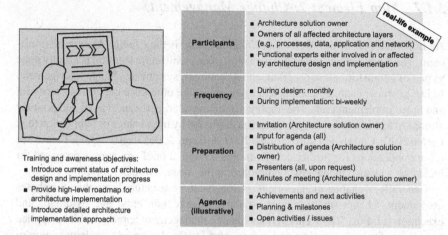

Training and awareness objectives:
- Introduce current status of architecture design and implementation progress
- Provide high-level roadmap for architecture implementation
- Introduce detailed architecture implementation approach

Fig. 5.15 Exemplary training preparation elements

Table 5.6 Summary of training and awareness element

Map Element 6: Training and Awareness	
Objective	Plan, prepare and conduct trainings for various target groups that will be either involved in or affected by the later architecture implementation
Prerequisites	• The work on selected elements of the Architecture Solution Map (e.g., strategic alignment, architecture framework, architecture ownership, architecture design, change management, or architecture implementation) has already commenced • Top management support of training and awareness sessions
Main activities	• Elaboration of training plans and preparation of training concepts related to architecture solution development and implementation • Identification of target groups that should participate in the trainings • Conducting of training sessions and awareness events in parallel with the execution of other Architecture Solution Map elements
Key results	• Training and awareness sessions for selected architecture solution elements executed • Internal communication triggered with the objective to inform stakeholders and staff about the endeavor of planning and implementing an architecture solution

Table 5.6 summarizes the most important aspects of the *Training and Awareness* element of the *Architecture Solution Map*.

5.1.7 Map Element 7: Change Management

An important part of the *Architecture Solution Map* is the *Change Management* element. The development and implementation of an architecture solution is usually done as part of a transformation program, and it requires a systematic change management. In fact, a proper change management might make the difference between a successful and failed implementation. The importance of considering the individual and political attitude in a structured change approach is discussed by broad range of publication (e.g. Carter 2013; Cummings and Worley 2009; George 2006; Kotter 2007). Please see those publications for further details. In this section, a brief structure is proposed for including a change management approach in the *Architecture Solution Map*.

Generally, a change management approach can be structured into the four phases assessment, change management strategy, change concepts and roadmap, and execution (cf. Fig. 5.16). This general change management approach can be applied to any transformation project and, hence, also for the design and implementation of an architecture solution.

The assessment phase (i.e., phase 1 of the change management approach) focuses on the identification and analysis of change needs. A first change impact analysis and also a diagnosis of change maturity and transformation culture of the telecommunications operators are performed. Possible results of the assessment phase are a change impact matrix, stakeholder engagement, and a manual for change and communication concepts.

In phase 2 of the change management approach, the change management strategy is defined. As part of the change management strategy, a description of clear change goals is elaborated. In this respect, it is important that the top management of the telecommunications operator is willing to drive and to support the

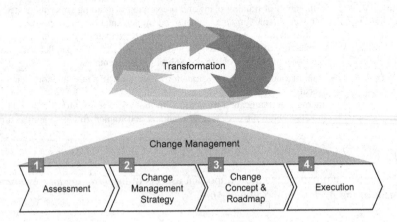

Fig. 5.16 Change management approach[5]

---

[5]Based on the topical development results of Detecon's change management team.

Table 5.7 Summary of change management element

Map Element 7: Change Management	
Objective	Development and application of a change management approach as part of a transformation program with the objective to plan and implement an architecture solution
Prerequisites	• The planning and implementation of the architecture solution is part of a transformation program and the need for change management activities have been identified by key stakeholders • The work on selected elements of the Architecture Solution Map has already commenced
Main activities	• Development and realization of a change management approach consisting of the four phases assessment, change management strategy, change concept and roadmap, and execution • Conducting of regular meetings with key stakeholders and affected target groups • Identification of risks and critical issues • Development and implemented of countermeasures for identified risks and issues
Key results	• Change management approach developed and implemented in the organization as part of the transformation program • Countermeasures for identified risks and issues identified as well as implemented where feasible

change goals and the change strategy. Through the top management commitment, a critical mass of change supporters can be recruited in the organization. This is essential for a smooth transformation and respectively for the successful development and implementation of the architecture solution.

A detailed change concept and roadmap is elaborated in phase 3 of the change management approach. Necessary change actions are defined that serve as input for the change roadmap. The change roadmap itself includes change actions, timelines, milestones, resource estimations and responsibilities; hence, a detailed specification of the planned change activities.

The execution of the change activities and the implementation of the change roadmap are part of phase 4 of the change management approach. In this phase, regular meetings have to be performed in order to monitor the progress, the efficiency and the effectiveness of the change activities. Risks and critical issues have to be identified and countermeasures have to be implemented.

Table 5.7 summarizes the most important aspects of the *Change Management* element of the *Architecture Solution Map*.

5.1.8 Map Element 8: Architecture Implementation

Based on the detailed design of the different architecture layers the implementation of the architecture solution has to be planned and executed. For this the *Architecture Solution Map* foresees a dedicated *Architecture Implementation*

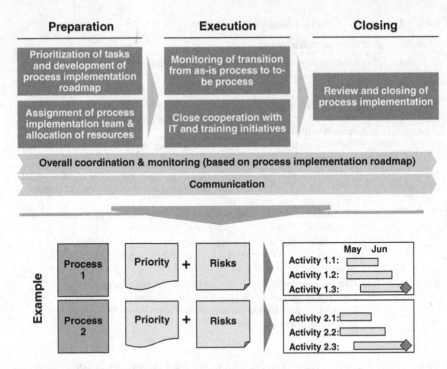

Fig. 5.17 Process implementation approach

element. Based on relevant project experience, it is recommended not to underestimate the complexity and the effort of implementing the architecture solution.

First, a structured implementation plan combining implementation tasks, responsibilities, dependencies and constraints is essential. The effort for the implementation can range from almost zero for the documentation of the as-is situation to complex IT changes for an overall reengineering. Usually the implementation approach consists of a preparation, execution and closing phase. In the preparation phase, the implementation tasks are prioritized and an implementation roadmap is developed. During the execution phase, the transition from the as-is state to the to-be state is monitored, where a close cooperation with the IT department and training initiatives are required. The closing phase has the objective to review the implementation and to derive countermeasures if the implementation did not lead to the desired outcome. Figure 5.17 illustrates the implementation approach and an example for a high-level implementation roadmap. Both are derived from a real-life project example and based on a process-driven approach.

Furthermore, the definition of clear responsibilities for each implementation task is necessary. In this context, a clear differentiation of responsibilities in the program organization and responsibilities in the line organization is essential. Typically, the implementation provides the hand-over from the program organization to the day-to-day business performed by the line organization. Figure 5.18 provides a real-life example for process implementation activities and highlights the responsibilities of the process owners, sub-process partners and others.

Task	Business Process Office	Process Owner	Sub-Process Partner	Others
Implementation				
Plan deployment and assign implementation team	Follow-up and documentation	Commitment of dates and responsibility for implementation requirements	Support – if required	IT: Provision of necessary IT requirements
Communicate and publish new process design	Official communication and publication	Forwarding of communication to all required employees	Support – if required	-
Conduct deployment (e.g. IT implementation, training, etc.)	Follow-up, monitoring and escalation (if required)	Overall responsibility for implementation	Support – if required	Support of line organization according to their responsibility (e.g. IT, HR)
Start of operations	Monitoring	Overall responsibility	Operational responsibility	Line organization: execution

Fig. 5.18 Process implementation tasks and responsibilities

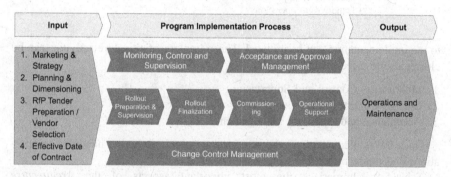

Fig. 5.19 Technical implementation tasks

Architectural implementations require intensive technological changes (e.g., roll-out of network elements, new software systems). In this case, the implementation typically includes the involvement of suppliers. Based on the functional and technical requirements, a bidding procedure and selection of suppliers is performed. Furthermore, an overall management and monitoring of the suppliers are necessary. In Fig. 5.19, a structure for technical implementation tasks is proposed.

Typical technical implementation endeavors require input parameters such as marketing and strategy information, planning and dimensioning data, RfP (Request for Proposal) documents and vendor selection decisions. Finally, a signed contract is needed between the telecommunications operator and the supplier responsible for providing the technical infrastructure (hardware and software), and for rolling out the infrastructure. Once the input parameters are available, the core activities related

Table 5.8 Summary of architecture implementation element

Map Element 8: Architecture Implementation	
Objective	Development and execution of a structured plan for implementing the different architecture layers while taking resource and time constraints into consideration
Prerequisites	• Detailed design results for the different layers of the architecture solution are available • Design results are agreed amongst the key stakeholders and top management representatives of the telecommunications operator • Allocation of sufficient resources and staff to the implementation activities • Strong involvement of architecture layer owners
Main activities	• Development of a structured implementation plan combining obvious dependencies and constraints • Elaboration of an implementation approach consisting of a preparation, execution and closing phase for each architecture solution layer • Close cooperation between architecture layer owners and representatives from involved departments • Definition of clear roles and responsibilities for the implementation of the architecture solution layers
Key results	• Detailed design results related to the different architecture solution layers successfully implemented • Handover to operations

to the implementation can be initiated. Those implementation core activities can be clustered into three distinct phases:

1. Monitoring, Control and Supervision;
2. Change Control Management; and
3. Acceptance and Approval Management.

Once the implementation core activities are accomplished, a handover to operations and maintenance takes place.

Table 5.8 summarizes the most important aspects of the *Architecture Implementation* element of the *Architecture Solution Map*.

5.2 Transformation Types and Organizational Responsibility

Transformation endeavors in the telecommunications industry can be categorized into three different types:

1. Transformations with a *strategic* orientation
2. Transformations with a *technical* orientation
3. Transformations with an *operational* orientation

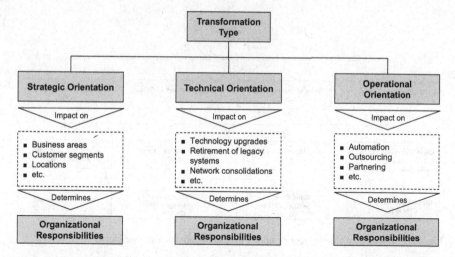

Fig. 5.20 Transformation types and functional orientation

Transformations with a strategic orientation can have an impact on the business areas, the customer segments, or the locations of the telecommunications operator. Transformations with a technical orientation, however, might only focus specifically on technology upgrades, retirement of legacy systems, or network consolidations. Finally, those transformations that have an operational orientation have the objective to increase automation, to outsource certain business functions or to establish partnerships with external parties for service fulfillment (cf. Fig. 5.20).

For each transformation type, different top management stakeholders of the telecommunications operator have to be involved. Therefore, this section focuses on the exemplified description of top management organizational responsibilities for each of the three transformation types (i.e., strategic, technical and operational). This is facilitated through the definition of one concrete transformation endeavor (i.e., one concrete instance) as examples for each transformation type.

Before assigning organizational responsibilities to a transformation endeavor, the objectives of the transformation have to be formulated and agreed amongst the top management representatives of the telecommunications operator. The objectives and also the nature of the transformation do influence the determination of organizational responsibilities. Therefore, it is important to briefly outline the most common top management positions amongst international telecommunications operators and their respective functional focus areas (cf. Fig. 5.21).

The Chief Officers illustrated in Fig. 5.21 usually have the following functional roles in the organization of the telecommunications operator:

- **CEO**: Corporate officer in charge of managing the telecommunications operator with the objective to maximize the value of the entity.
- **CSO**: Responsible for defining and implementing the strategy of the telecommunications operator in close cooperation with other entities of the organization.
- **CMO**: Development and realization of the telecommunications operator's marketing strategy including the responsibility for product development and launch management.

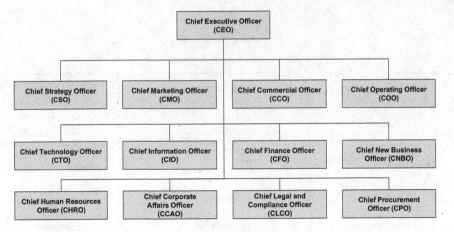

Fig. 5.21 Exemplary chief officer positions in telecommunications operators

- **CCO**: Planning and execution of sales activities by using various sales channels across the telecommunications operator's organization.
- **COO**: Responsible for the management and steering of the whole organization and of all business processes in the organization of the telecommunications operator.
- **CTO**: Development and realization of the telecommunications operator's technology strategy by utilizing latest technology trends across the industry.
- **CIO**: Responsible for defining and realizing the IT strategy of the telecommunications operator by taking existing legacy systems and constraints into consideration.
- **CFO**: Management of all finance-related matters within the telecommunications operator including budgeting, financial management, controlling and billing.
- **CNBO**: Identification and establishment of new business areas with innovation and growth potential besides the traditional telecommunications business.
- **CHRO**: Management of all personnel-related matters in the organization of the telecommunications operator including recruitment, personnel development, skill management, staff trainings, payroll, etc.
- **CCAO**: Responsible for a broad range of topics including e.g., investments, wholesale, international carrier relations, regulatory affairs, risk management, and external affairs.
- **CLCO**: Responsible for legal matters in the organization of the telecommunications operator including the definition and implementation of as well as the adherence to compliance policies and regulations.
- **CPO**: Focusing on sourcing, procurement, and supply management activities for the telecommunications operator with the objective to achieve synergies of scale.

A concrete organizational structure does not necessarily include all the above chief officer positions. In practice, different positions might be merged—for example, CSO and CMO. Furthermore, besides a pure functional structure, also market-related or product-related criteria could be used for defining chief officer position. In this case, the single responsibilities are divided between different positions below the chief officer level (e.g., Director or Senior Vice President).

The following presents a detailed discussion of three concrete transformation endeavors for describing the suggested organizational responsibilities at the top management level.

Example 1—Strategic Transformation

A telecommunications operator is currently offering mobile services to consumers. In the future, the telecommunications operator intends to also provide fixed broadband services to consumers through the existing fixed network of a competitor who has to open its network as obliged by the regulatory authority. The shift from providing pure mobile services to providing broadband services has to be planned from a strategic perspective and affects various units in the organization including, but not limited to, the technology department, the marketing and product department, the sales department, the regulatory department and even the recruitment department for the recruitment of fixed broadband experts.

Example 2—Technical Transformation

The demand and need for higher bandwidth is constantly increasing and telecommunications operators constantly have to compare the bandwidth of the existing network (fixed or mobile) with the future demand. There are many examples where telecommunications operators have already upgraded their mobile network from 3G to 4G technology. This technical transformation requires a strong involvement of the technical department of a telecommunications operator that is responsible for network planning, network roll-out and also network quality assurance.

Example 3—Operational Transformation

Through the penetration of IT systems in organizations, the number of process steps that are automated and executed by IT systems is constantly increasing. Automated processes promise to have the advantage of being executed more efficiently and effectively. In this respect, the automation of processes through the implementation of a new CRM system should be considered as one example. In order to achieve this, an operational transformation of the organization towards process automation has to be carefully planned and requires close coordination between the department responsible for Business Process Management (BPM) and the IT unit responsible for implementing the CRM system.

The three transformation examples provided above have different objectives, and therefore different top management stakeholders in the organization have to be involved in the transformation. A critical success factor for a transformation project is a strong top management sponsorship and commitment. Based on the above exemplary chief officer positions, the following top management representatives are linked to the three transformation examples:

Strategic Transformation Responsibility

For the strategic transformation from pure mobile services to fixed broadband services while using the existing fixed network of a competitor, it is recommended that the Chief Strategy Officer (CSO) has the overall responsibility—at least at the beginning of the transformation. The strategy transformation example provided also requires the involvement of additional top management representatives including the Chief Technology Officer (CTO), the Chief Marketing Officer (CMO), and the Chief Commercial Officer (CCO). The offering of fixed broadband services also requires the involvement of the regulatory department. At a later stage and ideally starting with the implementation of the strategy, the overall responsibility should be transferred from the CSO to a commercial top management representative—for example, the CCO to emphasize the importance of successfully launching new fixed broadband products and services.

Technical Transformation Responsibility

This transformation example to migrate a mobile network from 3G to 4G technology is technology driven and hence the CTO should carry the overall responsibility. The migration of a network from 3G to 4G technology requires a solid technical understanding. An early involvement of the commercial department is important because the propositions that can be provided via a 4G network will differ from the products and services currently provided over the 3G network. The commercial department has to be involved well in advance in order to be able to enhance the product portfolio and roadmap accordingly.

Operational Transformation Responsibility

The operational transformation focuses on the automation of processes related to the implementation of a new CRM system and requires an intensive collaboration of the business process department and the IT department. The business process department is located in the division of the Chief Operating Officer (COO). In this transformation example, telecommunications operators should appoint the COO and the Chief Information Officer (CIO) for jointly taking over the transformation responsibility. Through appointing the COO and the CIO as representatives for the transformation project, the close interaction and cooperation between both units are ensured.

Ideally there should always be one dedicated sponsor who has the overall accountability for the transformation project. The project sponsor has the task of overseeing the project, including project structure, progress, risks and critical issues, results, and project success. On the project execution level, most transformation projects require a cross-functional involvement of several organizational units. Cross-functional means that stakeholders from various departments that have a certain functional orientation—for example, strategy, marketing, technology or IT —actively contribute to the transformation.

5.3 Transformation Project Examples

Having introduced general transformation types and related organizational responsibilities in Sect. 5.2, this section focuses on providing concrete project examples that are related to the transformation of telecommunications operators. All project examples are real-life examples from the telecommunications industry. The objectives of a transformation project have an impact on the project scope, the project activities, the project results, the architecture elements in focus, and also the critical success factors of the transformation project. This section provides a structured description of ten different transformation project examples, out of which four concrete transformation projects are described as detailed example cases in Sect. 5.4.

Transformation Project Examples

1. Process and Quality Diagnostic Study
2. Business Process Management Establishment
3. Organization Restructuring towards Customer Centricity
4. Post Merger Integration (Organization, Processes and IT)
5. Introduction of a Next Generation (NG) Mobile Network Technology
6. Strategic Business Division Establishment
7. Introduction of an OSS—Customer Orientation (cf. also detailed example cases description in Sect. 5.4.1)
8. Introduction of an OSS—NGN-based (cf. also detailed example cases description in Sect. 5.4.2)
9. Introduction of a CRM System (cf. also detailed example case description in Sect. 5.4.3)
10. Introduction of Process Architecture (cf. also detailed example case description in Sect. 5.4.4)

The selected transformation project examples in this section (cf. Tables 5.9, 5.10, 5.11, 5.12, 5.13, 5.14, 5.15, 5.16, 5.17 and 5.18) and the detailed case study descriptions provided in Sect. 5.4 are based on real-life examples in the telecommunications industry. The practical experience with different kinds of transformation projects confirms that a special emphasis should be put on the critical success factors before project start and during project execution. The early involvement of project stakeholders from different departments and the communication towards affected employees about the project objectives are two key suggestions that might influence the success.

Table 5.9 Process and quality diagnostic study

Example 1: Process and Quality Diagnostic Study	
Project description	Analysis of the current situation with respect to the existence of a Business Process Management (BPM) and quality assurance function
Project activities	• Definition, review and agreement of study criteria • Review of existing BPM and quality assurance function • Review business process framework and processes • Consolidation of study results
Project results	• Agreed set of criteria for review and development of recommendations • Documentation of BPM and quality assurance review results • List of major BPM and quality-related pain points • List of improvement measures with the objective to close existing gaps • Final report summarizing all project results
Architecture elements in focus	• Processes • Organization (due to linkage of processes to organization)
Critical success factors	• Early involvement of key project stakeholders and communication measures • Usage of reference models and industry standards for analyzing the existing Business Process Management and quality assurance function

Table 5.10 BPM establishment

Example 2: Business Process Management (BPM) Establishment	
Project description	Development and implementation of an end-to-end Business Process Management function (i.e., an organizational entity)
Project activities	• Review of existing BPM activities • Identification of existing gaps and pain points • Development of high-level BPM guidelines • Detailed conception and elaboration of BPM guidelines (including, e.g., templates, roles, responsibilities) • Development of a roadmap for implementing the BPM function

Project results	• Approved BPM guidelines consisting of major BPM modules including:
	  1. Business Process Framework
	  2. Process Ownership Model
	  3. Process Design Methodology
	  4. Training Concept
	  5. Process Performance Management System
	  6. Process Implementation Concept; and
	  7. Continuous Improvement Process
	• Implementation roadmap/guideline

Architecture elements in focus	• Processes • Organization
Critical success factors	• Top management involvement and commitment • Early agreement on high-level BPM guidelines amongst stakeholders • Usage of reference models and industry standards • Communication and trainings

Table 5.11 Organization restructuring towards customer centricity

Example 3: Organization Restructuring towards Customer Centricity	
Project description	Analysis and transformation of the existing organizational structure with the objective to be more customer centric
Project activities	• Define strategic goals of the company for being more customer-centric • Review of the existing organizational structure with respect to customer centricity • Development of different scenarios for target organization • Scenario evaluation according to predefined criteria and final selection of one organizational scenario • Perform detailed organizational design
Project results	• Defined company strategy in terms of customer centricity • Transparency about and evaluation of current organizational structure • Different organizational scenarios • Final selection of one organizational scenario • Detailed design of target organizational structure
Architecture elements in focus	• Strategy • Organization
Critical success factors	• Consideration of company strategy and existing organizational structure • Continuous change management to involve employees that are affected by organizational restructuring • Agreement on evaluation criteria for organizational scenarios amongst key stakeholders • Top management commitment, involvement and decision making

Table 5.12 Post merger integration (Organization, Processes and IT)

Example 4: Post Merger Integration (Organization, Processes and IT)	
Project description	Plan and perform Post Merger Integration (PMI) for the organization, processes and IT landscape of two telecommunications operators
Project activities	• Review, assessment and evaluation of organization, processes and IT landscape of both telecommunications operators by considering PMI rationale and strategy • Identification of organization, process and IT elements relevant for the newly integrated company • Development of a consolidated target organization, process and IT landscape • Elimination of redundancies
Project results	• Transparency about existing organization, processes and IT evaluated with respect to PMI • Selected organization, process and IT elements relevant for integrated company • Detailed target picture for organization, processes and IT • Implementation plan and roadmap for organization, processes and IT

(continued)

Table 5.12 (continued)

Example 4: Post Merger Integration (Organization, Processes and IT)	
Architecture elements in focus	• Strategy • Organization • Processes • Applications
Critical success factors	• Consideration of the strategy of the company that will be newly integrated • Manage expectations of key stakeholders in both organizations • Early and effective communication towards affected employees • Fast decision-making processes and top management involvement

Table 5.13 Introduction of a Next Generation (NG) mobile network technology

Example 5: Introduction of a Next Generation (NG) Mobile Network Technology	
Project description	Upgrade of the existing mobile network to a Next Generation Network Technology
Project activities	• Analysis and evaluation of the performance and capacity of the existing mobile network • Determine current and future capacity demand for mobile network • Perform mobile network planning based on capacity requirements and latest technology trends • Vendor selection for the provision of hardware and installation • Network roll-out
Project results	• Performance overview of existing mobile network • Overview of capacity requirements and demand forecast • Mobile network architecture and detailed planning • Selected vendor for mobile network hardware and roll out • Network roll-out and migration
Architecture elements in focus	• Applications • Network Infrastructure
Critical success factors	• Consideration of demand forecast and latest mobile technology trends • Selection of experienced technology vendor • Overarching program and vendor management to ensure successful mobile network roll-out and migration

Table 5.14 Strategic business division establishment

Example 6: Strategic Business Division Establishment	
Project description	Establishment of a strategic business division with the objective to provide products and services in vertical markets besides the classical telecommunications business
Project activities	• Formulate mission, vision and strategy of new business division to be established for specific vertical markets • Define product and service portfolio as well as roadmap of strategic business division • Define detailed organization structure, skill profiles, processes and IT landscape for strategic business division • Implementation of organization, processes and IT
Project results	• Strategy of new strategic business division defined • Product and service portfolio and roadmap finalized • Target organization, processes and IT landscape defined • Implementation concept for organization, processes and IT
Architecture elements in focus	• Strategy • Organization • Processes • Applications
Critical success factors	• Strategy of new business division aligned with key stakeholders • Organization, processes and IT have to support the product and service offering • Overarching program management to handle different implementation streams and their complexity

Table 5.15 Introduction of an OSS (customer orientation)

Example 7: Introduction of an OSS (Customer Orientation)	
Project description	Development of a concept for introducing an OSS with the objective to achieve a stronger customer orientation, efficiency increase and the improvement of competitiveness
Project activities	• Analysis of existing processes, applications and data • Development of processes of the customer-centric, technology and product domain following a top-down approach • Usage of TOGAF for developing the Enterprise Architecture Framework • Mapping of processes and application functions for software solution selection • Development of data as an overarching architecture element
Project results	• Detailed process descriptions for the customer-centric, technology and product domain • Concepts that are compliant to leading industry standards such as TOGAF and TM Forum • Linkage of processes, data and applications
Architecture elements in focus	• Processes • Data • Applications

(continued)

Table 5.15 (continued)

Example 7: Introduction of an OSS (Customer Orientation)	
Critical success factors	• Develop consistent concept for the linkage of processes, data and applications • Usage of leading industry standards for concept development and vendor selections

Table 5.16 Introduction of an OSS (NGN-based)

Example 8: Introduction of an OSS (NGN-based)	
Project description	Introduction of an OSS based on an NGN for the resolution of the existing complexity to enable the decoupling of production and product
Project activities	• Development of a complex architecture used for the bidding process to implement standard software • Execution of several trainings related to the introduction of TM Forum standards • Development of detailed process flows related to the technology domain following a top-down approach • Definition of business requirements based on the developed processes • Mapping of processes and application system groups
Project results	• Detailed architecture linking processes, data and applications • List of business requirements for the selection of standard software • Overarching and detailed data model
Architecture elements in focus	• Processes • Data • Applications
Critical success factors	• Develop consistent concept for the linkage of processes, data and applications • Usage of leading industry standards for concept development and vendor selections

Table 5.17 Introduction of a CRM System

Example 9: Introduction of a CRM System	
Project description	Introduction of a standardized Customer Relationship Management (CRM) system for several vertically integrated telecommunications operators that belong to the same group of companies
Project activities	• Detailed design of processes related to the customer-centric domain applying a top-down approach • Development of use cases related to the interaction with customers • Elaboration of an overarching data model • Linkage of processes, data and applications • Perform a vendor selection for the introduction of a CRM system based on standard software

(continued)

Table 5.17 (continued)

Example 9: Introduction of a CRM System	
Project results	• Concrete use cases supported through process descriptions of the customer-centric domain • Detailed concept for the linkage of processes, data and applications • List of functional and non-functional requirements needed for vendor selection • Selected vendor for CRM system implementation
Architecture elements in focus	• Processes • Data • Applications
Critical success factors	• Develop consistent concept for the linkage of processes, data and applications • Completeness for developed use cases • Detailed specification of functional and non-functional requirements for vendor selection

Table 5.18 Introduction of process architecture

Example 10: Introduction of Process Architecture	
Project description	Introduction of an overarching process architecture as basis for a company-wide management of business processes
Project activities	• Development and introduction of an overarching process architecture • Execution of trainings with strong TM Forum involvement • Reengineering of two reference process flows which are newly developed • Mapping of processes to application system functionalities • Derivation of requirements for process improvement based on the existing software systems
Project results	• End-to-end process architecture framework for customer-centric, network, customer, product and technology domain • Detailed reengineering for two selected reference process flows • Implementation support for two reference processes • Optimization of operational efficiency
Architecture elements in focus	• Processes • Data • Applications
Critical success factors	• Long-term implementation supported for newly developed reference processes • Involvement of various departments during process design and implementation to ensure cross-functional optimization

5.4   Detailed Example Cases

The reference architecture described in this book has been implemented in various real-life projects. While the previous section has described project examples, this section provides four detailed example cases of real-life projects.[6] According to the design science principles proposed by Hevner et al. (2004), the four case studies are an evaluation of the artifacts proposed in this book. *Case study 1* represents the comprehensive usage of the suggested reference architecture (cf. Chap. 4). Further restricted is *case study 2* that implements selected artifacts of the process layer, the application layer and the data layer. The focus of *case study 3* is an Operating Support System (OSS) in which the entire architecture solution is applied for realizing the OSS and also the identification and integration of related elements. *Case study 4* comprises the complete process layer as well as selected elements of the application and data layer.

5.4.1   Case Study 1—Introduction of an OSS (Customer Orientation)

The first case study shows the implementation of the reference architecture including the suggested strategic objectives.

This comprises the whole process layer, i.e., the domain structure, the reference process flows for the customer-centric domain, technology domain and product domain. Furthermore, on the application layer, the processes are linked to application functionalities, on the data layer an overarching data model is used, and the introduction of a Next Generation Network (NGN) is related to the network infrastructure layer.

The company for this case study is a vertically integrated telecommunications operator offering telecommunications products to residential and business customers. The offered telecommunications products include fixed and mobile telephony, Internet, IPTV and business solutions. The company held a monopoly for a long time, however is now facing competition through other telecommunications operators. The examined project has the objective to develop a concept for introducing an OSS and is part of a complex transformation program that, amongst other topics, includes the technical introduction of an NGN. From a strategic perspective, a stronger customer orientation, efficiency increase, introduction of new product bundles and an improvement of the competitiveness are desired.

---

[6]The project cases are anonymized but based on real-life projects conducted by the two authors.

The project is split into two phases: (1) as-is analysis and (2) target design. With regards to content, the focus lies on processes, applications and data. The network infrastructure is considered in another project of the transformation program and is based on an NGN.

TOGAF—the Enterprise Architecture Framework is used as method. The introduction of the target design includes comprehensive trainings that are performed in cooperation with the TM Forum. As part of the as-is analysis, the architecture solution described in this book (cf. Chap. 4) is used as reference in order to examine the current situation and to identify weaknesses. The target design follows a top-down approach. For the process architecture, the customer-centric domain is used as a starting point. In addition, a mapping of processes and application functions is performed that serves as input for the bidding of concrete software solutions. The developed concepts are discussed with the project stakeholders in workshops for final approval. At the same time, 28 specific requirements are documented for the customer-centric domain in order to customize the architecture of the historically grown structures.

After finalizing the customer centric domain, the product domain and the network domain are designed. Thereby the reference process flows are used as a basis, and a mapping to the application functionalities is performed. In total, 35 changes are documented on an operational level. As overarching architecture element the data model is designed. The focus lies on the flexible configuration of products consisting of services and resources.

Figure 5.22 shows exemplary extracts of the project documentation. The illustration top left shows the general structure and explains the usage of the introduced framework along the different layers for processes, applications, data and network. On the top right, the assignment of processes and applications for the reference process flows *complaint-to-solution*, *problem-to-solution* and *trouble ticket-to-solution* is shown exemplarily. According to the described artifacts, no information from the network domain is required for the *complaint-to-solution* process, and therefore the process can solely be executed in the customer-centric domain with access to the Business Support System (BSS). For the *problem-to-solution* process, however, access through the technology domain (via the *trouble ticket-to-solution* process) to the network layer is necessary. In this case, the BSS and OSS are accessed. The lower representation shows the correlation between processes and data. Through the end-to-end process *order-to-payment*, a product is requested by the client through an order. Technically, this is realized through the end-to-end process *production order-to-acceptance* in the technology domain that fragments the product into services and resources. After product provisioning, the relevant data for invoicing are consolidated in one invoice.

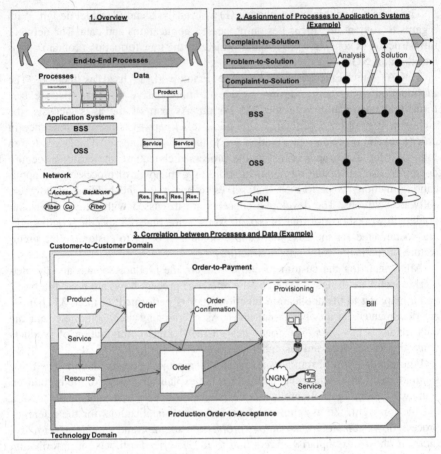

Fig. 5.22 Context and usability of artifacts

The examples clearly show the usage of the artifacts being introduced in this book. In this respect, the complexity of the concrete solutions becomes clear, and therefore only selected examples can be demonstrated in this section. Table 5.19 provides a summary of the case study.

Table 5.19 Introduction of an OSS (Customer Orientation)

Project scope	
Project	OSS introduction
Company	Vertically integrated telecommunications operator, former incumbent
Region	Asia
Customers	Residential and business customers
Products	Fixed, broadband, mobile, IPTV and business solutions
Architecture layers	
Strategy layer	Stronger customer orientation, efficiency increase, introduction of new bundle products, improvement of competitiveness
Process layer	Reference process flows for the customer centric domain, technology domain and product domain
Data layer	Overarching data model as integration element
Application layer	Assignment of processes to application functions as fundament for the bidding of concrete software solutions
Network layer	NGN introduction as part of an overall transformation
Evaluation	
Rationale for usage	Through the usage of artifacts many concepts can be reused whose development otherwise would have caused significantly higher project cost and a longer project duration.
Conclusion	The used artifacts were confirmed in workshops and, on an operational level, 63 changes for the adaptation of the historically grown structures were required.

5.4.2 Case Study 2—Introduction of an OSS (NGN-Based)

The third case study focuses on the implementation of the components of the architecture solution that are relevant for the OSS. In this case the reference architecture is used as fundamental structure. In detail, the reference process flows of the technology domain, the mapping to application systems for OSS, the overarching data model as integrating element, and an NGN are used. Furthermore, the correlation of the process layer and the application layer is used to identify and integrate the elements relevant for the OSS.

In the context of this case study, a vertically integrated telecommunications operator is considered that provides telecommunications products to residential and business customers. The offered telecommunications products include fixed telephony, Internet, mobile telephony, IPTV and business solutions. The objective of the project is the introduction of an OSS based on an NGN. A complex architecture is developed that is used for the bidding process to implement standard software. The existing software systems are historically grown and have vertical silos. For the resolution of the existing complexity, an overarching OSS will be introduced that enables the decoupling of production and product. The strategic objectives are efficiency increase, customer orientation, and the flexible integration of new partners as well as the realization of new products.

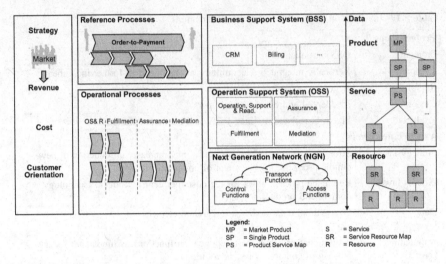

Fig. 5.23 Operating support system introduction

Besides conceptual development, the project also comprises several trainings. The usage of TM Forum standards as common terminology for the bidding processes is a must. The focus lies on the OSS of the application layer. The mapping between processes and application system groups and the reference process flows of the technology domain are used as the basis for defining the business requirements for the OSS. For further classification, however, a comparison with the BSS is required, which is performed considering the reference process flows of the customer-centric domain, customer domain, and product domain. In addition, an overarching and detailed data model is developed that is compliant with the data layer introduced in this book and to the reference data model SID developed by the TM Forum. An important element is, again, the flexible reproduction of products, services and resources.

For the distinct decoupling of sales and production as introduced and suggested in the architecture solution, a clear separation between the customer-centric and technology domain—and therefore also between BSS and OSS—is designed. Figure 5.23 illustrates the context through an extract of the project documentation.

The results of the case study are shown in Table 5.20. In summary, the artifacts presented in this book are used for the introduction of an OSS, and its implementability through the usage of standard software is confirmed.

Table 5.20 Introduction of an OSS (NGN-based)

Project scope	
Project	OSS introduction
Company	Vertically integrated telecommunications operator
Region	Europe
Customers	Residential and business customers
Products	Fixed telephony, broadband, mobile telephony, IPTV and business solutions
Architecture layers	
Strategy layer	Efficiency increase, customer orientation, flexible integration of new partners and realization of new products
Process layer	Reference process flows of the technology domain for the derivation of requirements, reference process flows of the customer-centric, customer and product domains for integration
Data layer	Overarching data model as integrating element
Application layer	Focus on OSS, usage of functions for structuring
Network layer	NGN introduction
Evaluation	
Rationale for usage	Through the usage of artifacts many concepts can be reused whose development otherwise would have caused significantly higher project cost and a longer project duration.
Conclusion	The implementability was confirmed through the realization of an OSS based on standard software.

5.4.3   Case Study 3—Introduction of a CRM System

The second case study shows the implementation of the following artifacts of the reference architecture: reference process flows for the customer-centric domain, mapping of processes to application functions, and the overarching data model as integrating element. The subject of the case study is several vertically integrated telecommunications operators based in different countries and belong to the same group of companies. The objective is the introduction of a standardized Customer Relationship Management (CRM) system that is realized through customized standard software as depicted in Fig. 5.24. In this instance, only residential customers are considered and the offered telecommunications products include fixed telephony, Internet and mobile business.

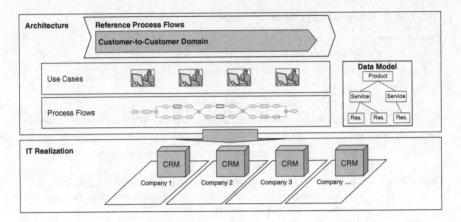

Fig. 5.24 CRM system introduction

From a strategic perspective, the focus lies on the reduction of IT costs through the introduction of standard software, increase of customer orientation and flexibility improvement. The objective of the project is the introduction of a CRM system based on standard software and, with respect to the reference architecture, the BSS of the application layer is in focus. For structuring the functional requirements, the reference process flows of the customer-centric domain, the respective detailed activities and the overarching data model are used. The compliance of the architecture solution with TM Forum standards is an important prerequisite for being able to realize a mapping towards standard software.

However, the TM Forum standards alone are not sufficient to achieve an end-to-end view of the business requirements. In this respect, the mapping between reference process flows and standard software is used for structuring, documenting and evaluating the business requirements. For that purpose, comprehensive workshops with the involved business units and the implementer of the standard software are conducted. In total, 2130 functional requirements are identified and all of them are assigned to the processes of the customer-centric domain. The described overall context is depicted in Fig. 5.24 using an exemplary extract of the project documentation.

In Table 5.21, the results of the case study are summarized. At the same time, the case study confirms a realization of the suggested artifacts for the customer-centric domain through CRM standard software in a complex project.

Table 5.21 Introduction of a CRM System

Project scope	
Project	Introduction of a CRM system
Company	Several vertically integrated telecommunications operators
Region	Europe
Customers	Residential customers
Products	Fixed telephony, broadband, mobile business
Architecture layers	
Strategy layer	Reduction of IT costs through the introduction of standard software, increase of customer orientation and improved flexibility
Process layer	Reference process flows for the customer-centric domain
Data layer	Overarching data model as integrating element
Application layer	Mapping of processes to application functionality and standard software for structuring, documentation and evaluation of requirements
Network layer	Not considered
Evaluation	
Rationale for usage	The architecture solution provides with the used artifacts the possibility to structure, document and evaluate the business requirements from an end-to-end perspective and enhances the standards of the TM Forum.
Conclusion	The used artifacts were confirmed in comprehensive workshops. In addition, through the mapping to standard software the implementability was demonstrated.

5.4.4 Case Study 4—Introduction of Process Architecture

The fourth case study includes the usage of the whole process layer. Furthermore, two reference process flows of the customer centric domain are implemented in IT systems with the objective to realize a process improvement. For this, the architecture solution as primary structure, the relationship between processes and application systems, and the overarching data model are used.

The subject of this case study is a vertically integrated telecommunications operator offering fixed telephony, Internet, mobile telephony, IPTV and business solutions to residential and business customers. The company held a monopoly for a long time; now, however, it is now facing competition through other telecommunications operators. The objective of the project is the introduction of an overarching process architecture as basis for a companywide management of business processes. The project is part of a transformation program with the strategic objectives to improve customer orientation as well as efficiency and flexibility. The functional focus of the project is mainly on the process layer. The reference process flows for all domains are used without any changes as basis for the process architecture. The compatibility with eTOM is a mandatory requirement.

During the introduction of the process architecture, comprehensive trainings with strong TM Forum involvement are provided. In the first step, the introduction of the process architecture is done based on existing systems and organization structure. The final acceptance is a result of workshops conducted with the involved process owners.

Two reference process flows of the process architecture undergo a process reengineering as part of a pilot, and are newly developed on an operational level. In this respect, the processes are mapped to the application system functionalities of the existing software systems. The requirements for process improvement are derived based on the existing software systems, whereas at the same time the developed concepts of decoupling products, services and resources as well as the overarching data model are used as guideline. The identified improvements are realized in the existing systems and the new processes are implemented through trainings on an operational level. For the new processes, an efficiency increase of 40 % is measured based on the cumulated working hours in the first two months after the realization.

Figure 5.25 shows an exemplary extract from the project documentation of the process architecture. A hierarchical structure for the process architecture is chosen that is structured into five domains on the highest level. All five domains are further specified in the project. In Fig. 5.25, the detailing of the customer-centric domain is presented that has been done according to the suggested reference process flows.

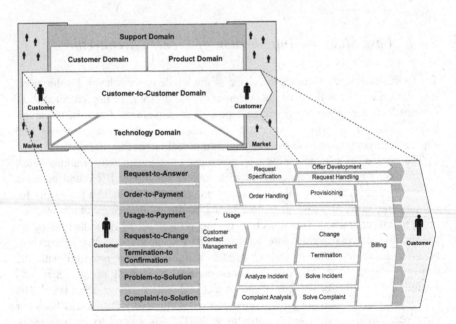

Fig. 5.25 Process architecture introduction

Table 5.22 Introduction of Process Architecture

Project scope	
Project	Introduction of a process architecture
Company	Vertically integrated telecommunications operator
Region	Asia
Customers	Residential and business customers
Products	Fixed telephony, broadband, mobile telephony, IPTV and business solutions
Architecture layers	
Strategy layer	Efficiency increase, customer orientation, flexibility
Process layer	Reference process flows of all five domains, piloting of process improvements for two reference process flows of the customer-centric domain
Data layer	Overarching data model as requirement for the overall process improvement
Application layer	Mapping of existing software systems through the application system functionalities for the two reference process flows of the pilot implementation
Network layer	Not considered
Evaluation	
Rationale for usage	The existence of a reference architecture for processes was the driving factor for the execution of the project. The compliance to eTOM was a mandatory requirement.
Conclusion	The entire process layer was implemented. A measurable efficiency increase of 40 % for two reference process flows of the customer-centric domain was realized through a pilot implementation.

The case study is described in Table 5.22. It can be summarized that, through the case study, the whole process layer for all domains is confirmed according to the recommended process flows in this book. In addition, the potential for enhanced efficiency increases by applying the suggested architecture realized through a pilot implementation is illustrated.

References

Ahlemann, F. (Ed.). (2012). *Strategic enterprise architecture management: Challenges, best practices, and future developments, management for professionals*. New York: Springer, Berlin.

Brown, J. T. (2008). *The handbook of program management: How to facilitate project success with optimal program management*. New York: McGraw-Hill.

Carter, L. (Ed.). (2013). *The change champion's field guide: Strategies and tools for leading change in your organization*, 2nd edn., and updated. San Francisco. Calif: Wiley.

Cummings, T. G., & Worley, C. G. (2009). *Organization development & change*, 9th ed. Australia, Mason, OH: South-Western/Cengage Learning.

Dinsmore, P. C., & Cabanis-Brewin, J. (2014). *The AMA handbook of project management*. New York: AMACOM.

George, W. (2006). Transformational leadership. In W. B. Rouse (Ed.), *Enterprise Transformation: Understanding and Enabling Fundamental Change* (pp. 69–78). Hoboken, N.J.: Wiley-Interscience.

Harvard Business Review Press. (2013). HBR's guide to project management.

Hevner, A. R., March, S. T., Park, J., & Ram, S. (2004). Design science in information systems research. *MIS Quarterly, 28,* 75–105.

Jones, G. R. (2013). *Organizational theory, design, and change* (7th ed.). Upper Saddle River, NJ: Pearson.

Koenigsaecker, G. (2013). *Leading the lean enterprise transformation* (2nd ed.). Boca Raton: CRC Press.

Kotter, J. P. (2007). Leading change: Why transformation efforts fail. *Harvard Business Review, 85,* 96.

Lankhorst, M. (Ed.). (2013). *Enterprise architecture at work: Modelling, communication and analysis,* 3rd edn. The Enterprise Engineering Series. Springer, Berlin.

Laudon, K. C., & Laudon, J. P. (2012). *Management information systems: Managing the digital firm* (12th ed.). Boston: Prentice Hall.

Van Den Berg, M., & Van Steenbergen, M. (2006). *Building an enterprise architecture practice.* Netherlands, Dordrecht: Springer.

Ward, J., & Peppard, J. (2002). *Strategic planning for information systems,* 3rd edn. Wiley series in information systems. Chichester, West Sussex, England, New York: Wiley.

Westland, J. (2007). The project management life cycle: a complete step-by-step methodology for initiating, planning, executing & closing a project successfully, Repr. ed. London [u.a.]: Kogan Page.

Wigand, R. T., Mertens, P., Bodendorf, F., König, W., Picot, A., & Schumann, M. (2003). *Introduction to business information systems.* New York: Springer, Berlin.

Index

© Springer International Publishing AG 2017
C. Czarnecki and C. Dietze, *Reference Architecture
for the Telecommunications Industry*, Progress in IS,
DOI 10.1007/978-3-319-46757-3

Printed in the United States
By Bookmasters